FROM PHYSICS TO FAITH

Mackenzie Neufeld

From Physics to Faith
Copyright © 2020 by Mackenzie Neufeld

All rights reserved. No part of this publication may be reproduced, distributed, or transmitted in any form or by any means, including photocopying, recording, or other electronic or mechanical methods, without the prior written permission of the author, except in the case of brief quotations embodied in critical reviews and certain other non-commercial uses permitted by copyright law.

Tellwell Talent
www.tellwell.ca

ISBN
978-0-2288-4827-1 (Hardcover)
978-0-2288-4826-4 (Paperback)
978-0-2288-4828-8 (eBook)

*This book is dedicated to Paula Paulgaard,
Bruce Dickie, Grant Gay, and Theresa Robinson.*

Ask, Seek, Knock

Matthew 7:7-8 – "Ask, and it will be given you; seek and you will find; knock, and it will be opened to you. For everyone who asks receives, and he who seeks finds, and to him who knocks it will be opened."

CONTENTS

Before We Begin 1
Welcome to From Physics to Faith 2
A Broken Past.. 4
Your Fight 33
Your Symphony... 36
The Beginning 39
From Physics to Faith 43
A 'Pandemic' of Questions 48
My Thoughts on 'Faith' 52
Love and its Beauty 56
The Cosmos of Our Minds 60
God and the Principle of Science................................ 64
Multidimensional Efficiency 67
The "Lateral Time Theory"... 70
Time Travel.. 75
'Debriefing' Time... 79
Energy and the Quantum—Part 1 86
Energy and the Quantum—Part 2 90
The Creation Story... 93
'Black Hole' Beauty 101
God's Creation Story.. 105
'Nothing versus Something' 109
Wrapping Up 112
Final Greetings... 115

BEFORE WE BEGIN

Welcome to From Physics to Faith—this is a book not like many others. Originally called Physics and Philosophy from a Sixteen-year-old, this collection of excerpts was my way to creatively piece together this 'puzzle of our universe.' My creativity, my imagination, and my passion were the fuel behind my endless ideas. However, the older I got, the more I realized it wasn't really 'physics' my mind was craving. Rather, it was my faith. It was my growing mind's recognition that there are things deeper than our human minds can comprehend, and I sought to have this connection. I sought to connect my physical, tangible, and earthly presence to a spirit and a reality deeper than myself. This is how I found my faith, and this is how I found God. Creating these ideas and writing this book was His plan for me, and following this path of passion and curiosity certainly has led me to Him. Hence, these ideas I had written as a young girl flourished into this book—a story of hardship, passion, advice, and spiritual growth. 'Knock,' and we will eventually find this truth, and this truth is known as the beauty of God. If we seek this beauty out, we will find it. It may take time, but God is always there to answer the beautiful door known as our faith. I introduce this book with something I wrote years ago. So, welcome, and enjoy this story—a story that is one to share.

WELCOME TO FROM PHYSICS TO FAITH

> *"As I am about to pour out my mind and my passion, I just want to point out one of my biggest insecurities about attempting to share my ideas. I am well aware that I am only a sixteen-year-old girl from a town most people haven't even heard of. I do not have a university degree, nor do I have a title, but unfortunately the reality is that in life you have to make your way from the bottom to the top."*

I am just a student who is still figuring out how to learn, yet I honestly believe that the most genuine of ideas stem from a mind of simplicity. Maybe my theories are not entirely 'scientifically accurate,' but that's okay. Every child, regardless of their education, experience, or knowledge, has something to offer, whether they know how to show it or not. I will admit that these 'theories' of mine may be beyond the empirically observable, yet when it comes to true theoretical physics that's all our mind has the capability of offering.

For years, I've lain awake at night with dreams of the cosmos and heavens above, and amongst these dreams stemmed passion—passion for answers and meaning in a world that frequently seems to lack such. Before I discovered the useful tool of the internet, and before

I started reading textbooks and novels on philosophy, cosmology, and physics, I had been struck with a really powerful feeling and desire to learn. I felt that something in my life was very incomplete, and I had come to the conclusion that I needed to fill this void.

I have always, for as long as I can remember, been fascinated with the fact that there is so much that humans don't know. There is so much for even our pristine and intellectual consciousness to fathom. However, we have done a pretty darn good job as a species of working around this impeccable barrier of human limitation. From the discovery of the atom, to the understanding of the fabric of space and time, humans have shown that *our brains not only have the power to observe reality, but to discover it.* This void that encompasses my brain will only ever become filled as I learn more about reality, our universe, and the human potential on the deepest possible level. Sometimes, however, this requires a bit of creativity on my part.

There will never be a shortage of things to learn, imagine, and discover, and I've realized that it is never too early to start. I've always had a really big imagination, and I later realized that this was helping me try and understand what I really wanted to know. I developed ideas that I never really knew how to describe, and only recently have I been trying to find an effective way to present them. Whether or not I have found an 'effective' way, I've certainly tried to write something.

A BROKEN PAST

I'm sitting down to write this section, and oddly enough it is one of the last things I have decided to add to this book. Never once did I think I would pour out my entire life story for any and all human beings to read. However, looking at my personal journey through life, I think it is necessary to include. Although I am only eighteen as I write this, I have a story to tell, and I hope by sharing this story I am able to inspire all those who care to read this book and all those who are struggling just like I had.

You may think this 'story' of mine is not an entirely happy one, judging by the title, and you are partially correct. It has definitely not been easy. However, amongst all the darkness, tears, and despair, there was always a little light. There was always a little light at the 'end of tunnel.' There were moments where it was so incredibly tiny and dim, but it was always there. I always trusted that this light would one day bring me peace and comfort, and I feel blessed to say that it has.

To be honest, I have no outline or plan whatsoever as to how I am going to write everything I hope to share, but maybe I will start with right now. It is November 18th, 2020, and I am sitting in my bed with candles burning in the corner of my room, and I am feeling quite at peace. This is a new feeling for me—it seemed that in my life, at least for as long as I can remember, there was always something that didn't sit right with me. There was always this feeling of being 'on edge.' I never knew why, but I always felt that there was this 'void' inside me: a void that never seemed to fill itself.

Before I continue, I want to mention something that might give some context to my struggles. On October 30th, 2020, not too long ago actually, I was picked up from my university residence by an ambulance and taken to the hospital emergency. I had not eaten for days and I was beside myself with anxiety and depression. I am not lying when I say that I felt completely disconnected from myself as a human being.

In fact, it had been that way for months. The night before I was taken to the hospital, I had been awake for forty-eight hours straight. It was early Thursday morning, and I remember walking out of my room at 4 a.m., feeling like a complete zombie. Before I left, I got dressed up in my nicest clothes, put in my diamond earrings, and curled by hair. Then, off I went. I walked all around campus through every corner and nook of the university without a clue where I was going or what I was doing.

There were still stars in the sky, and I remember looking up and feeling like these stars were the only little things to remind me I am a 'human being,' and that there are things so much deeper than all our struggles. These stars acted as a little glimmer of hope for just a second. For a moment, they acted as my 'little light.' I continued walking, however, still feeling like I was in some sort of trance. I remember finding a little garden in a part of the campus I had never been to. There was a fountain and a picnic table there, and it was surrounded by trees. I sat at the picnic table and closed my eyes, putting my head in my arms, and I slept. I slept for hours at this picnic table and woke up just to see the sun rising. It was a new day, and perhaps a fresh start.

That morning led to a day that was quite honestly the worst day of my entire life. I walked back to my residence still not entirely feeling like myself, and I joined my online meeting for morning prayer. However, I didn't retain anything from the readings or prayer that day. My eyes just felt so glossed over, and I felt so, so dead inside. This was definitely caused by the lack of food and fluids in my body, but I do feel like the root of it was something so much deeper. It felt as if all of my struggles, my terrorizing memories and traumatic elements of my past, were just all flooding back.

My campus minister, a beautiful soul I talk about later in this book, realized something wasn't right. So, she called me on the phone, and we talked for quite a while. I still had not eaten in days and I was physically shaking and crying because I was unable to even hold my phone to talk to her. I genuinely felt like I was going to die. I had never felt this way in my entire life, and it is something I hope no human will have to experience.

I remember falling asleep nearly every minute, it was so awful. I fell asleep, I woke up and was confused about where I was and honestly who I was, and then I fell asleep once again. It was such a terrorizing cycle. I just remember waking up and seeing three paramedics in my room, and I was not really scared, just confused. They took my vitals—I was extremely dehydrated—and they asked me a few questions. But before I knew it, I was carried out of my residence by stretcher and taken into an ambulance and hooked up to an IV to get all the fluids my body seriously needed.

After a little while on the IV, I started to realize where I was and why I was being taken to the hospital, and I started feeling a little more alive and 'human.' I started talking to one of the paramedics who was beside me in the back of the ambulance, and I clearly remember our conversation. I started talking to him about physics and how much I love science, and that I had been writing a book since I was sixteen. I told him all about my love for the theoretical and philosophical side of science, along with all the theories I had proudly created as a child. I felt so excited. Just talking to another human being after having been alone in my dorm for weeks without human contact lifted my spirits so greatly.

Shortly after, I got to the hospital and was wheeled into Emergency. I got a bed almost right away, and while still hooked up to the IV, I just slept. The nurses gave me heated blankets and I just slept. A few hours later, it was probably about 10 p.m., my campus minister came. We had only ever talked at morning prayer, Mass, and Alpha, and this was the first time I had seen her in person. As soon as I saw her, I felt so warm inside. I felt so safe.

I sat up in my hospital bed and she sat on the chair beside me and we just talked. We talked as if we had known each other for years. Just talking to her made such a difference to my day, and despite everything I had been through, I felt at peace.

About thirty minutes later, the psychiatrist came into my little room and said she wanted to talk to me. Actually, the first thing she said was—"Wow, you are such beautiful girl. You shouldn't be here." I actually get this a lot. I feel like I have put on a mask my entire life—I put on a smile and nice clothes and pull myself together in every social situation, and I feel like no one ever truly recognized how much I was actually struggling. I always kept it to myself, so her comment didn't even faze me.

We went to a little private room in the hospital, and the psychiatrist started asking me a bunch of questions—why I was at the hospital, what led me to have struggles serious enough to be taken to Emergency, and others. When thinking about these, I just broke down. I broke down like I never had before. The psychiatrist was a nice lady, but she spoke very harshly with a heavy accent and a tendency to interrupt me as I was unwillingly trying to share my story. That moment of talking about all my struggles to this psychiatrist was the most scared I had ever felt in my entire life. I was absolutely horrified in the most soul-piercing way possible. It was terrorizing.

I told her that I hadn't eaten in days, I struggled sleeping, and that I was going for so many walks so late at night. Sometimes I walked outside, and sometimes I used my key to get into the buildings—specifically the older part of the Psychology wing where I paced the halls, so incredibly 'zombie-like' for hours on end. I told her that I had not felt 'right' since March, the beginning of the COVID-19 pandemic. I told her that it felt as if there was a huge, huge void inside me that could never be filled.

I later found out that she diagnosed me with bipolar disorder on the spot (which I then found out I in fact didn't have). But she also told me I would have to go to another hospital for long-term stay. More specifically, I would have to go to a psychiatric hospital—a psych ward. I cannot tell you how terrified I was at this moment. I was an absolute mess inside and out.

She also told me I would have to "reconsider school" for this semester, and hearing this just destroyed my heart. School was my life, and learning was my passion above anything else. I felt enraged that she was trying to 'make me' drop out of school for the rest of the semester. Looking back, she was absolutely right, but at the time this was absolutely not what I needed to hear.

During this whole time, my campus minister was in the room with me. She was with me though all my tears and she sat with me in patience and calmness. She was my 'safe person' in the room. I was physically shaking, with my legs bouncing up and down uncontrollably fast, and I remember she put her hand on my knee. At this moment, I felt a tiny bit of warmth and peace inside. Looking back, I was so, so blessed to have her with me that evening. As soon as the psychiatrist left, I broke down in tears once again—heavy, heavy sobs and what felt like a waterfall of tears.

My campus minister gave me a hug, and I felt safer than ever in her arms. She gave me a hug filled with love, with care, and with compassion. She held me until I felt safe to let go. She was my little hero that night. I will forever and as long as I live remember this night and the love and care she gave me. Looking back, she was the most beautiful reflection of God I have ever seen in a person. She will forever be this special person in my life.

The next morning, I was taken to the Alberta Hospital just outside of Edmonton to stay in "Unit 12-A." When I got there, I didn't know how to feel. I tried so hard to put on my mask. I had my blazer on, and I cleaned up all the mascara that had run down my face. Most importantly, I used every ounce of effort to walk with confidence as the lead nurse gave me a tour of the ward.

She introduced me to one of the patients who was in the hallway, and the patient seemed terrified. The nurse told her I was a "new patient," and the girl's face immediately softened. She looked reassured as she mentioned, "I thought you were staff." I knew someone would say something like that. It honestly just justified how powerful that mask I had put on my entire life was. Despite how put together I appeared to be, I felt so, so broken inside. My insides felt shattered, but I kept on a pretty face with a glowing smile—I kept on this mask.

There were three empty beds in my room, and after my tour the first thing I did was crawl under my covers in my bed and weep. I was the first patient on my room—no one else had been admitted at this point—so I felt free to cry. I messaged my campus minister and told her I did not feel safe where I was. I told her that I did not feel safe at this hospital at all. I told her that I needed her, and that I could not be there anymore and most certainly not by myself. I remember her messaging me back, reassuring me that I was safe and although she wasn't physically with me, her heart was still with me.

She called me on the phone a moment later, and we prayed. We prayed to God and offered our intentions to Him, and she prayed for me. She told me that God will take my fear away, that He is more than happy to carry the burden of my struggles for me, all the while offering me the strength to get through them myself. She also told me that I am not alone through this. She told me that I have a whole army of people cheering me on.

I genuinely felt more at peace after this conversation. I slept, and truthfully, I slept in peace. I still had not eaten anything for a few days, and there was no chance at all that I was going to walk myself over to the cafeteria to eat with the other patients. When I wasn't asleep, I started to feel more and more terrified.

A little while after, I was joined in my room by another girl. Like me, she appeared quite young, and she had a gentle face and was soft-spoken. She had a smile, and she was very friendly. She was my first friend there—my first 'safe person' at the hospital. I didn't know her story and she didn't know mine, but I started to feel safe knowing that I had a 'teammate' to get through this journey at the hospital. I started going to dinner, and I started eating—a little bit at a time. They were baby steps, but steps to recovery nonetheless.

The more I got out of my room, the more I developed a connection to some of the other patients. I realized that all of these patients had a story—a grand and powerful story just like myself. Each patient had a story filled with tears, laughter, despair, passion, failure, and success. We were all unique little pieces in God's puzzle, and we weren't at the hospital because we were bad people. Rather, we were there to get the help and resources we needed to get back on our path. We were all beautiful, intelligent, and kind children of God, but what made us different than most is that we had a story to tell. We had such powerful stories.

A few days later, on Monday, I met my psychiatrist. He was a friendly man with a deep voice and caring eyes. He asked me questions, and we talked about my brain. However, it felt like the same sort of conversation I had had with the psychiatrist in emergency. It was a lot less scary of course, but it felt very shallow—very superficial. I talked about my struggles at the surface, but I didn't share my 'story.' I couldn't; it was something I had kept inside me for so many years and I just didn't know how to talk about it.

There was so much I felt like I needed to share, but I didn't feel brave enough to talk about it. They were things that had never left my soul, much less entered into another person's. I talked to one of my nurses about my dilemma—that I had a story to tell, a very long one, but that I didn't know how to tell it. We talked for quite a while, and somehow the conversation led to me telling him about my book, which at the time was called *Physics and Philosophy from a Sixteen-year-old*. I told him about my passion for writing and my passion for science, and he told me that maybe I should write about my story. He said that maybe it would help me share my story if I thought about it almost as a 'physics paper,' although rather than physics it would be evaluating my very own brain.

So, that is what I did. I wrote this paper, this 'story,' for hours and hours—my entire life story. I cried when I was done—not tears of sadness, but rather fulfilment. I wrote about all the things and more that I share in this book. Some things will remain between me and my psychiatrist, but a lot of these things I am comfortable sharing. After all, my only wish through writing is to help inspire others who are in a similar position as I was.

Growing up, I was very shy, and I was very self-conscious. I was very timid, and I struggled to make friends, especially in elementary school. Like most 'shy' students, I struggled with reading out loud, I struggled with raising my hand in class, and I really, really hated going out for recess. To me, it seemed like a complete and utter waste of time, and it forced an unwanted element of social interaction.

I also struggled very severely with school. Back in Grade 1, I remember having a very difficult time with my independence. I always clung onto my teacher, and I was not emotionally independent whatsoever. I struggled with learning on my own and I severely lacked the self-confidence to do the work that was put out in front of me. This was the case through my entire elementary experience, but it hit a peak in Grades 5 and 6.

I remember being forced to stay in at recess to get extra help with my writing. I of course didn't mind that at all as I would had rather been anywhere than outside at recess. Writing was a skill that I immensely lacked. I simply could not write—I stared at my paper and just could never get the right words out. I also struggled with math, so much so that it honestly seemed as if my teacher, my patient and kind teacher, was angry with me.

I just could not figure myself and my struggles out. I wasn't 'dumb', but I definitely was not good at school. It always felt so institutionalized and so impersonal. it always felt that we were there just to recite information and move on to another 'level' where we would have to recite more information—dry, lifeless information.

At the surface, I know I just seemed like an ordinary shy little girl who had her struggles at school but was still sweet and content with herself. However, inside there was something so much more. There was something so incredibly 'messed up' or 'broken' inside my mind, and that is how I thought of myself for years. I blamed myself for all of my struggles, not realizing that none of what I was going though was my fault. I was sick, and I was very sick, but I kept it inside. I kept it inside for so, so many years, and I didn't tell anyone. I got no help for what felt like this 'monster,' this 'demon' inside me that drove me to utter exhaustion at the end of each day.

Every minute of every day, my mind was constantly interrupted by very 'intrusive' and paralyzing thoughts. They were very scary and traumatizing thoughts that no child should have to think about, but no matter what I did, they wouldn't leave me. I called it my own 'living hell.' I've never told anyone what these thoughts were, because I can't, and I don't want to—you can't put things like that into words. It was so much more than a word or a construct of sentences; it was just pure terror flooding through my brain. These thoughts controlled my life. Quite honestly, they were probably the reason why I hated being at school and had such a difficult time making friends.

They controlled every and all of my daily activities. I remember being in the washroom washing my hands after school, and I had one of these 'thoughts.' It just flashed into my mind without consciousness, as they usually did, but this time it felt more paralyzing than ever. I felt compelled to move my arm up and down in the air, over and over again, until these thoughts went away. I felt like I was possessed, like I was a doll controlled by some evil 'puppeteer.'

I remember being in tears and fearful that I would never leave my bathroom, as I felt trapped raising my arm up and down in the air. I was literally locked in my bathroom with this feeling of terror for what felt like hours. Stuff like this happened multiple times a day, every single day for years. It was beyond awful. As a ten-year-old, I remember staying up hours into the night every single night, coaching and counselling myself around these thoughts. It was like a nightly ritual where I talked myself out of letting these thoughts control me.

It was so incredibly exhausting, and it was no wonder I couldn't focus or perform well in school—my little growing brain lacked so much peace and sleep. For years, during the time of my life when I should have been carefree and full of childlike energy, I was acting as my own counsellor. No child should ever have to go through what I did.

I hated my life and I wondered why my brain was the way it was. I remember crying out to God some nights before bed asking why my brain was like this. Why did I have to struggle the way I did? None of my classmates seemed to struggle the way I did. I was angry, but I was mostly upset. From such a young age I felt as if the only way I could escape this terror was through death, and I feared that I wouldn't be alive to see my teenage years. I prayed that one day, I would feel free—free from my own brain. I prayed that I could live the life I more than anything wanted to live.

At this age I remember being very into art, specifically drawing. I was very creative, and drawing was something I often felt at peace doing. I remember always finding myself drawing one particular thing. I always drew pictures of happy families, with happy and carefree children. I made up these families. I gave them a mom and a dad and many little girls and boys—happy and carefree girls and boys.

More specifically, I focused on making the daughter that I drew perfect—completely and utterly perfect. I drew her as a beautiful little girl with beautiful clothes, and I drew her at every age of her life from baby to adulthood. I also gave her a story of my dreams. I gave her a name, I chose her outfits, I made her a report card of perfect grades, and I designed her a bedroom. I remember cutting pictures of showrooms out of a Christmas catalogue and pretending it was her bedroom. I gave her the perfect friends, and I imagined her as a star athlete, a perfect musician, and the happiest girl one could be. I did this nearly every day. I gave this entirely made-up person a story, and I invested so much time into making her the person I always wanted to be.

This is a *generous* summary of some the struggles my mind had to go through as a child. There is so much more, more than I know I could write, but this was the basis of it all. I was constantly on edge and anxious, and I despised the life I had to live. It didn't feel fair at all, especially how I felt so alone through it all. I was a patient girl, and I was respectful. I never acted out, and I never companied about what I had to go through. So, for most of my life no one thought anything of it. To the world, I was just a normal girl.

One certain life event that sticks out to me as I write, and it was something that happened when I was twelve. It was during my summer of Grade 6, and it was the summer when my little brother got very sick. I remember this day, because it was a Saturday, and we were supposed to go on a camping trip. However, my parents cancelled it as it was raining, and the weather was expected to not be in our favour. Looking back, I know God was watching over my brother that day.

If we would have went camping—off the grid and out of cell range—my brother wouldn't have made it to the hospital, and he would have died. I know my mind blocked most of that day out, just because of how traumatizing it was, but what I do remember is him being taken to the hospital due to his blood sugar. It was extremely high, and if it were not treated that afternoon, he wouldn't have made it.

I just remember being left at home, for an eternity it seemed, as I waited in suspicion to hear an update on my brother. My parents were with him at the hospital, and it seemed as if I was never going to get updated. I was so incredibly terrified, and I remember melting into the ground in my brother's room, pounding my fists on the floor and sobbing. I just remember yelling at God and asking him why this had to happen.

I was confused as to why this happened to be part of his plan. I was so angry—it seemed as if everything that could go wrong in my young life was going wrong. I loved my brother. He had been with me through a lot:

through all of our parents' arguments and through all of the tension in my household. I was just so frustrated at why God had to let this happen.

Thankfully, my brother got the treatment and help he needed at the hospital, and about a week later he was all set to go back home. I later found out he was diagnosed with type 1 diabetes, which happens when your pancreas basically 'dies' and it fails to do its job of producing insulin. This completely transformed not only my brother's life, but the entire dynamic of my home. Like any parent would, my parents focused immensely on helping my brother with this new life. I had so much empathy for him and I felt awful that he had to go through this, and that it would be something he was stuck with for the rest of his life.

He also had many other health struggles, specifically with his autoimmune system. I was still quite little at the time, so I didn't entirely understand what he was going through. All I knew was that there was something serious, and that he would be requiring a lot of my parents' time and attention for a while.

I did understand this, but because I was still quite young, I felt really left out. I felt like I was constantly being left out of the loop about my brother's health, and because his health required my parents' focus and attention, I just didn't feel like I was part of the family anymore. I felt this way for years.

I began to feel guilty for having my own struggles. There was still no one I could reach out to for help with my brain. I felt like I had no right to feel bad about my mental health when comparing myself to my brother, but that didn't stop the problems in my brain from continuing to terrorize me. The only thing difference was that I felt much more alone. I was still acting as my own counsellor, but I lost all hope that it would ever get easier for me.

Looking back at this upsets me greatly. The fact that I felt like I had to go through this alone and was not given the support I needed during my entire childhood is what hurts me the most—even to this day. To me, the mental health of young children just seems so incredibly underlooked. We see children as happy, carefree little human beings with no problems in the world, but this could not be further from the truth. When the growing mind experiences struggles like my brain had, it is something that will affect an individual for the rest of their life.

I still do struggle with this aspect of my mental health, but it's not as extreme. As a young adult, I have learned to seek out my own help. I have learned to find those 'special people' in my life who can hold the flashlight for me while I'm lost on some of the darkest paths.

However, I am greatly affected by the memories of what should have been a 'happy childhood.' I was ashamed of my brain, and I never received the help or counselling I needed. If I were to change one thing about our world, it would be this.

Mental health is incredibly stigmatized, and it really should not be this way. It is just as serious as the physical health of a person, and there are ways to deal with it. It sometimes feels as if mental health is something both families and the healthcare system try to keep hidden. I know I am not the only child to have struggled the way I did, and it scares me that there are children in our world going through what I had to.

This is not right in the slightest, and for the rest of my life this is something I will pray for. Mental health is not something to be scared of. It is not something to fear and it is not something anyone should be ashamed of. If I could go back in time, I would tell 'little me' everything I just wrote. I would tell myself that I was valued, loved, and capable of so much more than my brain was telling me. I would tell myself that I was not broken; I was me, and despite my struggles I was a beautiful human being.

I talked a lot about my struggles, but out of all darkness comes light, so I want to share what helped me find this light. For me, things didn't start to get a little better until Grade 9—until I was about fourteen. Looking back, I am able to confidently say that this year was when I finally started to find my footing, and I am proud of myself that I made it that far to see the light.

Grade 9 was the year I started to find my passions, specifically my love for school. I developed such amazing relationships with my teachers who have acted as influential parental figures in my life. These relationships were the start of a 'new beginning' to my life. My teachers were those special people who reminded me that I have people on my team, and that I have a whole army of people to love and support me.

Just knowing this is so immensely powerful. To know you are loved is powerful beyond measure. I started to do well in school—very well, actually. My report card was golden. I also started to get serious about my love for running, and I had the confidence to join a track team. I won many races and I even placed fifth in the province of Alberta for the three-thousand-metre run. I also became serious about my 'career' as a little musician. I got many solos and was told that I played at a "very advanced level" for my age.

I finished off the year receiving a total of thirteen awards. I received class standing in every single one of my classes, I was honoured with awards for leadership, effort and achievement, athletics, the arts, and music. I was called "gorgeous" in the hallway by students I didn't even talk to, and I began to find my people—a group of beautiful girls I still feel honoured to call my friends to this day. *I became the person 'little me' had only dreamt of becoming: the girl in all my drawings.*

Now, this did not happen overnight. I still had—and I still have—my struggles, but there was something that led me to managing my ever-so-difficult brain. This something was science. This was the year I had found a passion for science—a science which was so much deeper than myself and my own life. This was the year I found my love for physics, and this was the year things started to look brighter for me.

It was the year I truly started to see the 'light at the end of the tunnel.' I began to ask questions that, looking back, were very advanced questions for someone at the age of fourteen to be asking. These questions were the beginning of my journey through science. My brain, I realized, had a strength, and that strength was creativity. This was and is my little superpower.

In pondering these questions, I started creating theories—explanations of phenomena at the farthest reaches of our cosmos. I did this every single night. Instead of talking myself out of my terrorizing thoughts, I created. I created, I wondered, and I painted this picture of our universe. I created a complex and colourful painting of our universe. My theories became something I was immensely proud of, and they gave me a sense of meaning. They were a reminder that I was valued. They were a reminder that my brain, contrary to what I had thought my entire life, was a beautiful thing.

I had so many ideas that when I was in Grade 11, I decided to write a book. I remember being so inspired by my biology teacher, Paula Paulgaard, and I wanted more than anything to show her all of the things my mind had come up with. This is natural, of course, to show someone you love and trust something you are so proud of, so that is what I did. I thought of ways I could share my ideas with her, and there just seemed so many of them to talk about.

So, I decided to write them down. That was the beginning of my book, and I called this book *Physics and Philosophy from a Sixteen-year-old*. It was the very original version of this book, where all of my original theories are discussed in great detail. I just remember that day in December when I came into her room, smiling from ear to ear as I told her that I had something to show her. I watched as she read the few pages I had given her, and I cannot even begin to describe the feeling I felt. It was such a beautiful feeling, and it was the beginning of my realization that I have a story to share with the world and that I would do anything in my power to share this story.

I once again began to see the light, and it was shining down so incredibly bright on my little life. However, a few months later this light went away for a little while. I was sixteen at the time, and I was part of a competitive badminton club where I trained multiple times a week with the same group of athletes. Some of us had been training together for years, so we did know each other quite well.

I remember about two and a half years before this, I was probably fourteen, I had met this girl. She was shy and quiet like myself, and we were both individuals who strived for perfection. I remember her always apologizing when she missed a shot in badminton, and that was something I had always done too. We were very similar in this aspect.

One day after practice, I remember noticing she had scars all over her arm, and as I myself have a history of mental health struggles, I felt like my heart sank into my stomach. I hated to see that another individual was going through this, especially this sweet girl I had the pleasure of calling my friend. I remember at the time my faith was not the strongest, but I did pray at times, and that night when I went home, I prayed to God to watch over her. I prayed that he would guide her to the little light that was shining for her at the end of the tunnel.

A few years later, back again when I was sixteen, I started to become quite good friends with this girl. At badminton we were always partners, and I always wanted to include her when I was with my other friends at practice. She was a quiet girl, but she had such a bright and beautiful soul.

There was one practice, however, when she wasn't there. I missed her, as I was missing my badminton partner, but I thought nothing of it. I thought that she might be sick with a cold, or that she might have too much homework, or that she was out of town. I didn't think too much about it. I just thought to myself, 'Oh well, I'll just see her next week.'

However, that weekend I received a message saying that she had died. She had killed herself. She was only a teenager, and I received a message letting me know that she had killed herself. I was speechless and emotionless. I did not know how to take in this information. Why? Why did this happen? This should have not happened to anyone, let alone someone that young.

My feeling of shock turned into anger, which turned into sadness, which turned into grief. I realized I could never talk to my friend ever again. I could never see her at practice, we could never finish our badminton game, and I would never see her smile or hear her laugh. But the thing that brought me the most pain was that I could never give her a hug. I could never give her a hug and tell her that I had been in her position—to tell her that her darkness would eventually turn to light. I could never save her, and that broke my heart. I cried myself to sleep every night for weeks, and my little 'physics' journey took a pause.

I had never really been religious throughout my life as my faith was always something that confused me, but after the death of my friend I began to find a little path to God. This is a path that my little feet are still leading me on. This is when I found myself walking over to Vice Principal Grant Gay's office one day after school.

Mr. Gay was and is an incredibly smart man. He was well versed in the sciences, and he was also a theologian. I knew he was a man of faith and that he had a very strong relationship with God, and this was something I began to crave. My physics had started to hit a wall after the death of my friend. I had started to realize there was something so much deeper than the physical laws and constructs of the universe, yet it wasn't until this period of grief and mourning when I realized this something was God. God, I began to understand, was this connection of the tangible to the intangible—the physical to the spiritual, and life to death. This is something I talk about in great detail later in this book. God was the only thing that could connect me to the friend I had lost. Love—God's love that He gave us —was the only thing that could transcend life itself.

This is what my Mr. Gay told me, and this is what stuck with me. I began to go to his office much more often, and we had wonderful conversations—the best kind of conversations. We had conversations about life, religion, philosophy, chemistry, and physics, and he even taught me how to file my taxes ☺. Through my relationship with Mr. Gay, I had started to find my footing with my faith. He was a mentor to me, and he led me to the path that was inside of me my entire life.

My relationship with God was always there, but Mr. Gay had helped me find it. He had helped me realize that the questions I was asking were not necessary questions of physics, but questions of faith. They were questions that the more I asked, the further and further I became from a potential answer. My questions led me to a wall that physics could not climb—it was a wall beyond all aspect of physical science. At this 'wall' was God. He was at the end of this path of questions my curiosity had led me on. This was a very monumental change in my life, and my relationship with God had helped heal the pain of missing my friend.

The remainder of my Grade 11 year definitely got better. I realized that my faith was the only thing that could help carry me forward in life, despite the pain of losing my friend. I got accepted into a national research program at the University of Alberta and I worked in a mechanical engineering lab with professors and grad students. Along the way, I had amazing conversations about physics, philosophy, and our universe. This inspired me to create and write my own ideas, and hence I worked my book every night. I was incredibly proud of all that I wrote. The summer of research and writing eventually led into fall and a brand-new school year—a fresh start. I was ecstatic to go back to school and tell my teachers all about my research experience. I was also especially excited to show them all the new pages I had written in my book.

The school year started out great. I excelled in all of my classes and I enjoyed everything I was learning; however, in October I started to have a lot of trouble with my learning. I started to struggle with things that I hadn't struggled with since elementary school. I struggled with my physics and math classes—classes that I loved so incredibly much. I also struggled with my writing, and it felt like I was my 'Grade 5 self' all over again.

I had ideas, that has never been a problem with my brain, but I struggled with putting them down on paper. I took hours and hours to do my homework after school, and it frequently felt as if I were re-teaching myself everything from my classes. I really struggled with keeping up in my classes and it seemed like I could not process any of the information I was learning.

It was absolutely awful, and I really did not want things to go back to how they were in elementary school. It got to the point where I was supposed to write an in-class essay for my AP English class, and I didn't write a single word on the page the entire hour. I started crying. Thankfully, I was writing in a classroom by myself—it was a snow day, and I was the only student who showed up. Thank goodness, too—I am a mess when I cry.

I guess what I am trying to say is that things were not easy for me in school. In fact, the more that I thought about it, the more I realized that they had never been—not even when I received all of my awards throughout junior high and high school. I had always struggled with learning and not feeling 'at the same speed' as my classmates. However, the only difference was that I stopped blaming myself.

I had always done the extra work and taught myself the information I struggled to learn in a regular class environment, and I did it without complaining. This was part of 'little me' that had still not really changed. I blamed myself for my struggles and did not realize that they were not my own fault.

It wasn't until that year when I academically 'crashed.' I remember being in my physics teacher Bruce Dickie's classroom one day after school, and I just could not keep myself together. I was always very good at putting on a mask and pretending that my life was perfect, but Mr. Dickie was someone who had coached me and taught me for years, so he was someone I trusted to see my struggles.

I remember crying and my mascara running down my face, and I remember ranting to him about how much I hate how 'institutionalized' school is. How there is so much more to learning than tests, and marks, and assigning a number value to a student's potential and success. I hated this part of modern schools, and I ranted the heck to him.

He was so very patient with me, and he just listened. He stayed with me for hours after school until I was calm and felt okay enough to walk home. I cannot tell you how much I appreciate him being there for me in that moment, and Mr. Dickie, if you are reading this, I really hope you know how much I value you in my life.

Shortly after, I was referred to my school's learning strategies advisor to talk about the struggles I had been facing with school not only recently, but throughout my entire life. I always had done well in school, so no one knew I had as much trouble with learning as I did. My advisor recommended that I see a learning specialist, so that is what I did. I underwent psychological evaluation and assessment, and that winter I found out I had a learning disability which had been affecting my 'processing speed.' In fact, it was pretty severe in how much it impacted my ability to learn all of these years.

Because I had put on this 'mask of perfection' my entire life, both my mental health struggles and my learning disability went unattended for years. If I am being honest, it has only been within these past few weeks that I truly recognize the importance of being honest with ourselves, especially with being true to our emotions—we are human beings, and there is a reason why we have them. Although this sounds incredibly cliché, it is so important. Realizing this has been a major step in my mental health journey, and it was a step that has helped lead me to recovery.

Another thing I would like to mention when trying to express my life story and struggles is this ongoing pandemic—the infamous COVID-19 pandemic. Back in March, I remember that finding out my school was going to be closed for the rest of the year felt like the worst thing I had ever heard. I just happened so suddenly.

I had such powerful and special relationships with all of my teachers, and being torn away from them hurt more than anything. School was my 'safe place,' and they were my 'safe people.' It did not bother me that I would be missing out on science labs, band concerts, track meets, or even my Grade 12 graduation, but what really affected me was knowing I wouldn't be able to see all the people that mattered so much to me.

This terrified me, and for months I became depressed. It was a different kind of struggle my mind was facing. I had never felt 'depressed' before. I had felt sad, overwhelmed, anxious, and trapped by my mental health struggles as a little girl, but never depressed. For months my brain was not healthy, and that lasted through my move to Edmonton as I began university.

I missed my teachers more than anything, and once again there was this void inside me that I craved to fill. I craved emotional affection and love, and I craved for someone to wrap me in their arms and tell me everything would be okay. My teachers had filled this void for me, and I didn't know how I could fill it on my own. I craved to give all of these people that meant more than the world to me a big hug. It simply broke my heart not being able to do this, and from March to November I felt this way. And now, here I am—in the middle of November 2020, writing my story.

After my conversation with the nurse in the hospital, I took his advice, and I wrote my psychiatrist a paper explaining this story. It was a long paper, but I shared everything I needed to share. I never got to print it out at the hospital. Instead, I remember walking into the little room where I met with my psychiatrist and I gave him my computer to read what I had wrote. He had his assistant with him, the one who writes all the notes about patients, so he had the computer read my paper out loud.

Hearing my entire life story be read out loud was so painful, but it was so incredibly relieving. Hearing at least ten years of struggles echo out of the computer and into the ears of a professional who could finally help me was the most relieving sensation I have ever felt. This was the moment. This was the moment where my struggles were heard. I was listened to and my years of pain and adversity were recognized. I broke down in tears as I heard my story being told aloud.

I had never shared those parts of my life with any other soul before, and I almost felt like a disappointment to the little girl inside me who had worked her entire life to keep on this mask. I remember going over to the sanctuary, the little private room in the hospital, and talking to my campus minister on the phone. I told her about the paper I had wrote and how much of my story I had shared, and I again broke down.

I felt as if the void inside me was only getting larger by sharing my entire life story with a doctor I had only just met. Although it was necessary, I really struggled with coming to terms that I had struggled as much as I had as a child. At that moment, I felt like that little girl who was scared of the power of her own mind. I was scared of the terror it caused her, and I was scared that her mask would slip—that the world would see who she really was. My mask had slipped that day; in fact, it had fallen completely off.

The doctors knew my entire story and were quite shocked that I had been as strong as I had as a little girl years ago. I told my campus minister that I felt empty—that I just wanted someone to hug me, to care for me, and to tell me "Good job; you made it this far." I longed for this recognition and physical protection more than ever. Despite sharing my story, I felt more alone and emptier than I ever had before.

There was something my campus minister told me that stuck with me, and I think about it whenever this feeling of 'emptiness' comes back my way. She told me to close my eyes and picture myself sitting in God's lap. She told me I was still God's little child—that He had been watching over me and protecting me all along. Although I am eighteen, she told me God exists to hold us. God exists to let us put our head on His shoulder and to let Him wrap His arms around our shaking or shivering bodies. This is what I did, and I instantly felt consumed with warmth. My heart rate slowed, and I felt more relaxed than ever. It was like I was giving someone the biggest and warmest hug ever.

That is how I felt, and at that moment, I felt truly safe. I will forever remember this phone conversation I had with my campus minister. She was my guardian angel those past few weeks, and she reminded me that God was and always is with me. He saw through my mask—He knew. This was all part of His journey for me.

I received names for all of the things I had struggled with—generalized anxiety disorder, major depressive disorder, obsessive compulsive disorder, and ADHD. But most importantly, I was able to feel comfortable with taking off the mask I had put on my entire life. I finally felt safe to do that, and this little 'hug from God' was my sign that I did the right thing.

This is my story, and this is the story I hope to share with the world. It is a story of how my love for physics helped me get through a lot of my biggest struggles, and how this love for physics turned into finding my faith. My relationship with God is as strong as it has ever been, and every single moment of my life—every moment of pain, passion, perseverance, and despair—had led me to this.

This is a book where I have attempted to share my story, but it is also a book where I have attempted to share my creativity with the world. This is something 'little me' would have wanted me to do. The rest of this book showcases my years of ideas: the ideas that have formulated theories upon theories during the peaks of my creativity. These are theoretical constructs of our universe—simply ideas that I have tried to fit together like a puzzle. For each of these ideas I attempt to connect a component of faith, and for me, this was the easy part. After all, I always say that there is no physics without faith. ☺

YOUR FIGHT

I find that when my thoughts become incredibly jumbled and my mind ceases to avoid this hurricane of clutter, I resort to writing. Writing is my safe haven. It's what I do when my bothersome mind refuses to let me take a breath. At the moment, it would be an understatement to say that I am confused about a lot.

To be honest, I am always trying to figure out why I'm here. What benefit do I myself contribute to this world? What can I offer? I am unable to find comfort in any answer I attempt to give myself. In fact, most of the time, I feel utterly pointless. I feel that my presence is a burden, and my aspirations are nothing but unreasonable fantasies of my mind. It is easy to put on a smile, to act confident, and to appear that you have your life together. Yet in reality, I'm not the person I'd like to think I am. I'm happy, but sometimes I wonder why.

One moment I find joy in the most subtle blessings of life, and the next I feel like a little child wanting to curl up in their mother's arms and seek protection from the hardships of life. One moment I have all the confidence in the world, and my plans for my future all lined up; however, a moment later I feel stripped of all of my faith in myself. I am so incredibly confused by the powers of the human brain—how my perspective of myself and the world around me can be so drastically changed in the blink of an eye. Why? Why am I like this?

I desire to be a source of light for the world: someone people can go to for inspiration, for advice, for love, and for friendship. However, how am I supposed to be the light for others when I can't even light up my own darkness? This deeply troubles me. One of my mentors once told me that when you are having a difficult time overcoming your own darkness, help someone. It's not about proving anything; rather, it is about reminding yourself that you're strong enough to fight for others when you can't fight for yourself.

It's amazing how well this works. Despite my crippling lack of self-confidence, helping others is the most rewarding sensation imaginable. It is such a special and selfless thing that allows you to truly recognize your potential as a human being. However, this is easier said than done. When I am with the people I love, I feel genuinely happy inside.

Yet attached to this feeling is fear that I won't be the perfect friend, the perfect daughter, or the perfect student. I fear that I will say the wrong thing or cause more harm than good. I fear that I am unable to be a source of light for the people that need me. This fear touches me, and it's incredibly hard to get over.

Although these footprints of fear may be invisible to others, they constantly trail my path. They tend to climb deeper into my mind, settling into my subconscious roots. It is inescapable. Perhaps this is the cause for anxiety, paranoia, and depression within the human race. It's hard to be strong, and maybe deep down this is why I originally attempted to write a book—to prove to myself that I have potential in this world. Although I may not be perfect, I have value—contrary to what my mind tends to tell me. Regardless of where my writing takes me, it is doing its job in allowing me to recognize this.

Now, I find that my brain dives readily into a philosophical evaluation and exploration of the human mind. Those close to me say that I give good advice. Perhaps it's thanks to this trait of mine, but maybe it is just because I needed to hear the same thing. However, sometimes there is only so much that words can do. Sure, they help, but until you truly understand them your internal light will cease to shine.

What I'd ask of all human beings, of every individual reading this book, is to not give up on yourself. Life will be difficult—trust me, it has been for me at times—but stick with it. There is a sun behind the clouds, and it will come out. Just give yourself time. Help others, but more importantly take time to look after yourself and find your light, because only you can do that.

YOUR SYMPHONY

Before I get into the depth of my ideas, I would like to quickly write to you: the reader. I would like to offer a little bit of advice, perhaps. If you are reading this book, I assume you are much like myself in that you have a passion for the philosophical and faith-based approach to understanding science. Perhaps you are a scientist with a strong faith, or perhaps without. Or perhaps you are not a scientist at all. I can say with confidence that I have been all of the above at one point or another in my life. Regardless, this book has something to offer you. It is a book which will make you think, and it is a book which I hope will inspire every soul that cares to read it.

Maybe you will relate to parts of my 'story,' or maybe you won't, and that's okay. However, if your mind is like mine, you may relate to setting the bar high. You attempt to exceed all expectations set in front of you and your mind will only let you rest when you do so. It's quite exhausting at times, but your drive and your passion are what keep you going.

Yet as a high achiever, it is inevitable that you will fall victim to your mind's persistent distractions of self-doubt. In fact, more often than not you will let this trait set you back. Now, this may seem very cliché, but I truly deem the topic of self-confidence crucial to bring up in the world of scientific academia.

There will be theories that you develop that ultimately bear no element of truth and simply deny all laws of physics; however, your limited knowledge refuses to make you aware of such. You will have ideas that you expect are original, and then you turn around to find out someone else had already brought your idea to the world fifty years ago. There will also be instances when you pour your heart out to developing your theory only to have someone else tell you it is missing a crucial and 'fundamental entity.'

To blatantly put it out there: these instances are inescapable. When you enter into the world of science, you are signing up for all of these instances and more. All that you can do as a scientist is persist. As a young and aspiring physicist, there are countless occasions when I hit roadblocks with my ideas and find myself in these situations. It would be a lie to say that I just shrug these situations off. It's incredibly debilitating. You feel as if your voice has no place in the world—that your contribution of passion is irrelevant. But let me tell you it's not.

Yes, some of us have quite minimal knowledge compared to some academics; however, this is what gives us our power and our light. Creativity stems from simplicity, and without the pristine foundation of simplicity, creativity would cease to shed its elegance within the mind. Human advancement comes from the ability to foster ideas without the bias of current knowledge, and 'information,' as useful as it is, exists as no more than a filler. Information is something for your brain to find comfort in when attempting to grasp onto familiarity. Don't get me wrong: it is crucial that we do 'step on the shoulders' of those who lived before us. It would be ignorant to ignore the art painted by these exceptional minds.

However, the flames of knowledge need to be tended to. The fire will lose its spark if we simply just observe it. Our minds must tend to it. It is our responsibility to add logs to this fire, not just rearrange those already burning. Regardless of how much you know, value the simplicity you possess because this has the potential to create something new. When this is mixed with the power of inspiration, your mind can do wonders. It can achieve the unachievable.

Appreciate what you can offer. Your ideas may follow those created before you, but don't be afraid to step into your own mind. Those who do may just unleash some of the pieces needed to complete the universe's greatest puzzle.

Now, give yourself time. If you expect yourself to get everything done before you turn eighteen, you are going to be unsuccessful. There is no rush; the universe is not giving out maps the second you take a break from your expedition. Use time to your advantage. Most importantly, however, please remember that the instrument of your mind will produce harmonies that with time will perform a symphony.

THE BEGINNING

> *"I've only been on this earth for sixteen-and-a-half years. I am not lying or being 'humble' when I say I don't know much, especially since I didn't really begin to question my surroundings until I was old enough to understand really how big the world is. Like every child, you go through a time in your life when you begin to observe your surroundings and start to question them. Why is the sky blue? Why can't my dog talk? How can Santa deliver all of the Christmas presents in one night? These are all questions a child will ask the same way an intellectual individual will begin to question the nature of their existence."*

Curiosity drives us all, the same way now as it did when we were children. I don't have the same experience as a sixty-year-old physics professor at Oxford, but I can guarantee you that I have the same extent of *passion*. Because of my lack of experience, or lack of years to ponder the 'mysteries of our universe,' I am forced to be creative with my approach to find any answers, and in order to find any answers you need to come up with questions.

I am not very familiar with many scientific equations, nor do I know what theories have or have not been proven true as laws. In other words, I have very little background information to answer the complex questions my mind is desperate to know. Because of this, I tend to create my own answers, not knowing if they are completely false and defy the 'laws of physics.' As long as my mind is questioning things that may not even have answers, I will always strive to create solutions, or 'theories.' Regardless of whether or not these make complete sense, they will be used as a foundation for further imagination.

In my opinion, creating a theory is like writing a story. It all starts with a spark of imagination, which leads to the creation of ideas, which all evolve to build off of each other and support further ideas. So, my curiosity and imagination are the only tools that have been helpful in somewhat answering the questions that keep me up at night.

You may wonder why I am so persistent on picking apart every little detail of the simplest and most obscure aspects of life. One may think this is a complete waste of time, and I understand why someone would think that. However, every philosopher and physicist you ask will tell you their own version of "because I'm curious." It is human tendency to be curious. Whether this curiosity becomes 'activated' is up to the individual. Regardless, it is a part of all individuals that have the ability to think for themselves. Whether you decide to flow with time or dissect it to pieces is up to you.

Before I jump right into the meat of my questions and ideas, I would like to bring up an important and underlying concept to existence: the idea of time. School starts at 8:35 a.m., lunch needs to be eaten at 12:00 p.m., track practice starts at 4:00 p.m.... Why does the concept of 'time' have such prominent control of our lives? To start with the typical philosophical question, we should ask what time *really* is. What does it entail, and what is its importance?

Whenever someone mentions time, I think of a schedule: a timetable of organization. The clock—a pristine tool used to break our day up into a series of numbers. Hours are composed of minutes, each minute is composed of seconds, and so on. Without time, there wouldn't be order, and without order there would be chaos. Hence, civilization has found a way to schedule human activity in an *orderly fashion*.

In the span of human existence, we have always managed to find a way of 'order.' In fact, every entity in existence has ultimately found order from chaos. From the physical world of atoms and the orbit of planets around stars, to the philosophical world of human thought and the recognition of our consciousness—everything knows what to do and how to do it. You don't see an electron reading an 'instruction manual' on how to behave; it simply does what it does as 'satisfaction' is reached when this sense of order is fulfilled.

At the same time, you don't need to think about how to think, as by doing this, you are thinking. As humans, our 'order' may be summed up as our consciousness: an intangible connection between our physical bodies and the world of our thoughts. However, what initiates this universal concept of order? Does everything happen at random, or do all events have a purpose, changing the future of the universe as a whole?

Now, time as a function of human organization is not '*true time*.' The universe could care less about our daily schedules, although both bear a sense of order. True time is a concept I created with the goal of uniting the past and the future of the universe to what I call the *energy reservoir*. I will not go deep into this concept in this segment as it requires quite a bit of precise detail in order to be fluently understood, but it will be mentioned later. However, to begin I would like to share a passage I wrote separate from this book. It is a passage that describes the beautiful moment when I had found faith from physics, and I think it is more than worthy of being included.

FROM PHYSICS TO FAITH

> *Dear Scientist and Doctor of the Church, natural science always led you to the higher science of God. Though you had an encyclopedic knowledge, it never made you proud, for you regarded it as a gift from God. Inspire scientists to use their gifts well in studying the wonders of creation, thus bettering the lot of the human race and rendering greater glory to God. Amen.*
>
> —*A Prayer for Scientists, Saint Albert the Great*

Like I mentioned previously, I have been all over the spectrum of science and faith. Yet ultimately, in the most fundamental sense they are one with each other. The more I thought about science, the more I was led to my faith, and the more I thought about my faith, the more I was led to science. It is truly a wonderful thing. I will attempt to explain this journey through writing to make sense of my thoughts in a way that my words seem to have trouble doing.

The past few weeks have been tremendously stressful, but stress not in the way I have normally experienced. I think as human beings, it is in our design to seek purpose and meaning for our lives, and regardless of what this may be, there is something special that all of us possesses. In each of us, there is this unique and personal little clock that gives us our tick. We don't simply live just to survive; we live to share this little part of us with the world.

As a little girl, I lay in bed with dreams of unlocking all the secrets of our universe. It was like a puzzle to me. All the pieces were there; they just needed to be put together. It was a very grand puzzle, but all the pieces were so simple and so organized. In fact, all of these little pieces were so incredibly fundamental. However, when you put them all together, the puzzle they produced was beyond beautiful. It was a piece of art that told a story not possessed by each individual piece of the puzzle. Only when these pieces were put together could they complement each other in a way they never could when apart.

This is what I thought of physics. These pieces occupied my mind every single moment of the day, and my little clock as a human being desired to find a way to join these puzzle pieces together. Since I was a little girl, I felt it was my calling to create this 'theory,' this one singular puzzle, that attached all of these pieces in physics together. I wanted to create a theory that described the origin of our universe and behaviour of the farthest reaches of our cosmos.

But the older I got, the more I started to wonder if my calling was actually what I thought it was. Was this puzzle—this theory of our universe—a puzzle of physics? Perhaps the world's definition of physics differed from mine. Perhaps 'physics' was just the name for something so much deeper, something that I loved so greatly. I was not sure anymore, and this is where my mind has been recently: in a state where I am trying to make sense of something that had made sense my entire life. Things have just not felt right.

What remains constant is this 'puzzle' binding together in my mind every day, every hour, and every minute. Perhaps this is not a puzzle of physics like I originally thought, but it is still there. This is a puzzle that connects me to something deeper than myself: so, so much deeper. It is something that has the power to connect the tangible to the intangible. Perhaps my connection to physics was a connection to God in the truest sense. Perhaps this puzzle forming in my mind was a puzzle of faith. The pieces were simple in their nature—as tangible and as fundamental as something could be.

However, when these pieces were put together, when they were connected, the result (or the 'theory' in the eyes of a physicist) beautifully explained all previous doubt. The 'gaps' in our universe became filled when the pieces were placed in the right place. Looking back, I realized this was 'faith' in the truest sense. When looking at these pieces individually, you would not think they encompass any sort of physical truth—maybe you will dismiss them or ignore them in that by first glance you will not see the truth and grandeur they possess.

However, through effort, through patience, through commitment, and through trust, these pieces will form the divine and ultimate truth, and this is faith. This is faith in the truest sense. To the little girl I look back on, this puzzle was one of physics, but now, I see that it is one of God. One of religion, of theology, and of the Holy Spirit.

This, to me, is the connection between science and theology. To the human mind they can easily be indistinguishable in their most fundamental sense, just as they had been for me for the longest time. They both serve as the bridge between the intangible and tangible world, and they both require the human being's capability of faith. Faith is our trust that these basic, simple puzzle pieces will form something so much greater than ourselves and our perceived understanding of the world.

I remember one night, not too long ago, sitting in my bed in tears and being overcome by immense grief and sorrow. I remember grabbing my rosary, clasping it in my palm and clenching it so incredibly tight. I felt my pulse beating in the centre of my palm, and I remember following along to the rhythm of my heartbeat penetrating though my tightly clenched hand. That night I realized something that brought me a lot of peace. I realized that in a way, we are little 'puzzle pieces' in a very unique sense.

Religion, theology, and faith itself are divinely beautiful puzzles formed by the union of their individual pieces: the union of us. We give life to faith—we are what unites the tangible to the intangible, and this is shown by God's presence in each and every one of us. Human beings are the body of faith—the 'puzzle pieces'—and when we come together through love, we form something so much greater than ourselves. We are God's creation, and through our union, one that requires faith, hope, and love, we form this 'puzzle.' We form God's beautiful and divine puzzle.

The pulse of my hand tightly holding this rosary that night reminded me that we are what gives life to faith. We are truly what gives faith its heartbeat. I think that as someone who has recently been struggling to find meaning and purpose in their life, realizing this gave me a lot of comfort. It gave me great perspective realizing that I am not alone in any journey life has for me. We are all God's puzzle pieces, and together with our faith we make up so much more than ourselves. This, more than anything, allowed me to recognize that my life does have a purpose, and through my profound connection to physics, I had found God.

As a little side note, I remember teaching my high school band teacher the link between physics, philosophy, and theology. I told him that I am a true believer that each one of these subjects ceases to exist on its own—they can only be *truly* understood when taken together. I remember telling him that physics is a science that creates explanations of our observable surroundings, philosophy is a field of study that recognizes there is a truth beyond these explanations, and theology and religion are a lifestyle that question why—why there is a truth, a reason, or a purpose in the first place.

A 'PANDEMIC' OF QUESTIONS

In observing and evaluating our society, I have noticed something. Through every relationship and conversation, I have realized how truly special our minds are. They are so very reflective and creative, yet so 'stumped' by our own inquiries. The mind of homo sapiens is so incredibly complex, yet so unknowingly paralyzed by an invisible barrier of limitations—a barrier that only our minds will build. Yet we have been designed to climb these intellectual barriers through the gift of curiosity.

Like I mentioned in a previous section, curiosity lights up so much more than the mind of a child. It is the fuel behind our mind's engine, and it is the very vehicle that drives our species to evolution. The universe doesn't care even the slightest about its past, future, or present state of affairs. Rather, this 'universe' is simply the means by which 'existence' is able to obey the laws set in front of it. It is our curiosity that poses these questions and theories as an attempt to explain this obedience in ways comprehensible for the likes of our capability to understand.

Regardless of the theories we illustrate or the equations we solve, we must humble ourselves upon understanding that we are constructing our own puzzle of reality. We have absolutely no credentials worthy enough of preaching laws that are beyond our mind's ability to prove. Our theories are theories: constructs of the mind designed to satisfy our thirst for meaning. However, the beauty of physics is that we don't have to be correct. We are paving our own pathway, answering to nothing but our mind's desire for understanding.

I am going to be completely honest; I don't know the 'right answers,' and I'm not going to pretend to. Yet what I can do is ask some questions that I am sure have intrigued more individuals than myself and share my attempts throughout the years to answer them. Before I begin to tackle many of the big questions flooding not only my mind, but plenty of the minds of those immersed in the world of academia, it is important that I address what these questions are.

To start on the foundation of 'time,' it is important to give ourselves some credit. We do have an idea what this pretentious concept entails; however, in our modern world more than ever, the beauty of time is being run over by ignorant simplicity. We are losing the ability to even comprehend the true and grand nature of time. It is much more than a schedule or a series of unwanted alarm clocks. So, in asking the typical 'philosophical question,' what exactly is this construct of 'time,' or more specifically my concept of *true time*? How does the nature of time tie into the prestigious *order* of the universe, and how exactly is such becoming filled?

Secondly, considering its grand design, it is important to question why and how our universe started, current theories aside. In addition to this, will the universe end? If so, why? What is beyond an 'end'? Along with this, what is the composition of the fabric laying across our universe? If this sheet of spacetime is not a physical entity, how can our universe be so cohesive? In an arena that denies physicality, how are celestial bodies able to assert an effect on each other millions of kilometres away?

Adding to our questions even more, we are required to throw in the concept of expansion. Why would the universe feel the need to grow and stretch out in every direction, in every dimension? What is causing our universe to do so? Where is it going? What does "expansion" really mean, and how can this so-called intangible arena abide by such a physical concept? I had also briefly brought up the concept of an "energy reservoir," but what is this energy reservoir? Also, what does the notion of "energy" exactly refer to? How does this energy reservoir have anything to do with the past, present, and future of the universe?

Along with all of this, I always was perplexed as to how all subatomic particles are identical in nature, yet they are all separate entities. How are such things able to be virtually indistinguishable, yet each occupy their own existence? Another thing I have always wondered is what is it that determines the amount of matter occupying our universe? And will substance ever 'die out'? If so, what does this look like?

What about the chaos of 'quantum physics'? How can such a disordered system prevail in such an ordered universe? Also, are there more dimensions to space? If so, where are they? In fact, what do we really mean by "dimensions"? What about black holes: what are they? Why and when do they form? And time travel—what does this entail, and is it a possibility?

Well, there you have it—a sneak peek into my busy brain every night. However, I am sure these questions have kept, and will continue to keep, many intelligent thinkers on their toes. These are only a handful of the grand mysteries shed by our universe, yet I would say that these are some of the big ones. The theories and concepts I have created may just be pieces to a much bigger puzzle, but they are my attempt to answer some of these questions that have keep me up every night.

Regardless of how many theories mankind can piece together, we will never really know the truth behind some of these questions. Some things will always remain a mystery; not even the language of mathematics nor the elegance of the human conscience will shed light on the answers. That treasure is left for the peaceful nature of the universe and reality. That being said, we are left with our curiosity and creativity to tell this story. Our ideas may be the only foundation of truth available to us. We may never know if we are right; however, our imagination is the best place our species can go to attempt to find answers.

MY THOUGHTS ON 'FAITH'

In looking back at my entire life, I can confidently say that I have been all over the 'faith spectrum.' My faith was not something I always struggled with, but it was something I always doubted about myself. It was if anything I chose to believe in didn't sit right with me, almost like there was always a 'piece to the puzzle' I was missing. I was agnostic throughout my childhood, as faith was never really something I thought about as a kid, but for quite a while in my teenage years I became an 'atheist.' I have grown in my faith and relationship with God so, so much since that period of my life, but life is a journey. Life is about learning and growing as human beings, and I am so proud at where I currently am on my journey.

During that period of my teenage years, there was always something that did not sit right with me about believing in God. I always thought that religion blindly filled the gaps in our knowledge of nature with this concept of 'God' without looking at potential undiscovered scientific reasoning. This was my rationale for not believing in the existence of God, but like I mentioned, one's faith is a journey, and mine just happened to be all over the spectrum.

The older I got and the more I thought, the more I realized that religion was not at all 'filling the gaps' with God. The principle of religion, or theology, for that matter, does not strive to 'fill any gaps'; rather, it strives to question 'why.' Why are things the way they are, and why does nature exist the way it does? Theology does not replace science questions with a one-word answer of 'God'; rather, theology recognizes that there will be always something beyond a 'why' question.

In other words, theology and religion do not attempt to fill the gaps of science. They do not attempt to answer all the questions science is unable to. This is so far from the case. For all physical phenomena, science breaks it down. Science asks why certain 'events' in nature behave the way they do, it questions what the cause of such behaviour could be, and it identifies the physical and fundamental components of this behaviour. Science uses this 'why' question to lead one to answers, discoveries, and explanations of nature. However, the more 'why' questions science poses, the further away one will be from a 'physical' and 'concrete' explanation.

Ultimately, one will eventually find one's self asking the rhetorical questions of 'Why is there something rather than nothing?', or 'What is the reason behind the existence of the universe?' These questions are very open ended and are not questions of science at all. However, the fact that we are drawn to asking these questions—to always asking why—feels extraordinary to me.

We are the only species that gets this far on our 'why' journey. It is beyond basic curiosity; it is something that once we begin to ask, we will never stop. Our minds will never stop wondering what is beyond that of which we know—and this is what drives any scientist.

However, regardless of what technology we have or how advanced our species has become, we will always hit a wall with our 'why' questions. There will always be a part of nature, a fundamental or cosmological part of nature, that we will cease to understand, and it is foolish to think otherwise.

> *"Meanwhile every man remains to himself an unsolved puzzle, however obscurely he may perceive it. For on certain occasions no one can entirely escape the kind of self-questioning mentioned earlier, especially when life's major events take place. To this questioning only God fully and most provides an answer as He summons man to higher knowledge and humbler probing."*
>
> —*Gaudium et Spes, Pope Paul VI - December 7, 1965*

To me, this special part of nature is the realm of God. It is a realm of reality that exists yet is so far from our understanding. It is so far from our understanding, yet it is something that we realize, and it is something that we accept. We are a species that recognizes there is something so much grander and deeper than our own existence. It is something that we recognize, but it is something we will never reach. This is how I think of faith, and I am incredibly strong in my faith and in my relationship with God.

I recognize that He is this realm of beauty and truth so much deeper than us, but I also recognize this part of nature is not something myself, nor any other human being other than Christ will truly understand. We know there is a 'nature' beyond our own knowledge, but we know there comes a point where we just won't be able to 'dig any deeper.' And this is the limitation of being human. We are not God; we are human beings. We are designed to recognize God and accept Him as this 'undiscovered element of nature,' but we are also designed to never truly understand his grandeur.

This is my thought process, and as someone who struggled to accept religion into their life for so many years, this is what led me to come to a belief in God. I am a scientist, and I do value the contribution of theories and questions posed by the scientific theory, but I also recognize there is something so much deeper than what can physically be discovered, and I credit my faith for allowing me this realization. To me this realization was special. It was special in that I realized God Himself will only truly transcend His presence through the human being, not any other species. God is so much deeper than us, but He designed us with the tools and internal compass to find Him, and to find that which is so much deeper.

> *"The intellectual nature of the human person is perfected by wisdom and needs to be, for wisdom gently attracts the mind of man to a quest and a love for what is true and good. Steeped in wisdom, man passes through visible realities to those which are unseen."*
>
> —*Gaudium et Spes, Pope Paul VI - December 7, 1965*

LOVE AND ITS BEAUTY

The First Letter of John 4:13-16 - By this we know that we abide in him and he in us, because he has given us of his own Spirit. And we have seen and testify that the Father has sent his Son as the Savior of the world. Whoever confesses that Jesus is the Son of God, God abides in him, and he in God. So we know and believe the love God has for us. God is love, and he who abides in love abides in God, and God abides in him.

Months ago, I remember talking to my high school principal in his office one day after school, and there was something about our conversation that stuck out to me. This conversation is something I will always remember, as we talked about what it means to be human in the most genuine and authentic sense. We are who we are not because of our intellect, our social status, or even our profound curiosity, but because of the potential we have to love—to love like no other species can. To be honest, I never realized how true this really was until that day, or perhaps not even until now amidst all of my struggles.

"We are loved because God loved us first; we all participate in that, and that's what lets us love. Regardless of whatever befalls us in life, love endures through it all." This was what my principal told me, and it will forever be in my heart as a little reminder when I need it most. Love is something that is so much deeper than it is quite often understood to be. It is what has the power to connect the tangible to the intangible, the physical to the spiritual, and life to death. 'Love' is why we pray, and it exists as this intangible connection to the beautiful presence of God's kingdom.

Love is why we read God's words through the Bible—the words only infiltrate to places deep in our soul when read with love. And lastly, love is the seed of our faith, a seed that when planted with care grows hope and passion and the true beauty of our own being.

Back when I was sixteen, I started writing my book with a simple idea I called the 'Cause, Effect, and Transition' principle where I talked about what I called the 'Rhythm of Order.' I elaborated on this more in the context of physics, but like I always say, physics and theology are one with each other. To put it simply, this is what I originally wrote:

The beauty of nature is that everything that happens ends. If it doesn't end, then it's still in the process of 'happening,' but if something starts—it will end. The reason that anything happens at all is to achieve its end, its final result or conclusion. I called the original event the 'cause,' and the end of this 'cause' I called the 'effect.' The order was what I referred to as the beautiful and eloquent process between a 'cause and an effect.'

The end of a 'cause,' or the 'effect,' has achieved this order, and what is between the cause and effect is what creates this order—the 'transition.' The 'end' is the result of a completed order, and as soon as what needs to be completed is completed, the beauty of the 'effect' is set forth. I like to think that everything has a purpose, every physical or cosmological event, but also every relationship, every moment of passion, and even every moment of hardship and despair.

In talking about physical phenomena, I always asked myself how is it that nature just 'knows' what events need to happen? Or how does nature know what events are required to be completed? I answered this by coming to the conclusion that things exist solely through their *potential* relative to where they are placed in a situation. The transition from 'start' to 'end,' or 'cause' to 'effect,' is the key to maintaining order.

Regardless of where something is going or what it is doing it will eventually reach a point of completion—and this is determined by its *potential*. When writing the introduction to my book I went on to talk about this section in the context of 'expansion'—otherwise known as the constant inflation of our universe. However, I truly believe that for all physical phenomena there is a spiritual connection. So, this morning I closed my eyes and let my mind wander into this topic.

I started to realize that 'love' itself exists as the element of 'order' I mentioned above. We are all physical beings, physical in flesh but bound by a spirit that cannot be explained by any physical law. Rather than a 'cause' of an 'event' like I used to explain nature's physical and cosmological phenomena, we are 'beings' that exist through the union of both our physical and spiritual presence.

However, only though love itself will we reach the spiritual—the 'intangible' element of our existence. *Love is this 'transition phase' of our existence, this beautiful and eloquent concept of 'order' I mentioned above.* Our physical being is in union with God only through genuine and authentic love, and this is what it means to be a *human being*.

When looking at Jesus, we know that he was a man entirely free of sin. He participated in love in the most genuine and authentic sense. He was and is the only model of human perfection. He was the son of God, man, but filled by the Holy Spirit—a spirit that exists solely as pure love. This love, the Holy Spirit, connects 'man' to God. It connects the Son, to God the Father. Love itself is this intangible connection to our creator and the Father of all nature.

To strive for love is something in our very own design as human beings, and Jesus acts as the perfect role model of this. He had the most genuine and authentic relationship with God his Father, and this relationship was through the Holy Spirit through love.

We will never be perfect human beings, but through love we are able to reach our highest potential as a species. We will reach the places our footprints cease to travel, and we will truly be able to share our gifts and talents with the world. Love is once again this 'transition phase' of our existence. Only love, the Holy Spirit, can guide us to God and to an order of life so, so much larger than ourselves.

> *"In His goodness and wisdom God chose to reveal Himself and to make known to us the hidden purpose of His will (see Eph. 1:9) by which through Christ, the Word made flesh, man might in the Holy Spirit have access to the Father and come share in the divine nature." (see Eph. 2:18; 2 Peter 1:4)*
>
> *—Dei Verbum, Pope Paul VI - November 18, 1965*

THE COSMOS OF OUR MINDS

Now, let's introduce some 'science.' To begin, let's start by discussing the most fundamental ingredients of our cosmos. Pretend the universe is like a computer screen. You as the observer have complete control of a little 'magnifying glass'; every click of your mouse leads your perspective further into the depths of the universe. You are viewing the universe around you consumed by its infinite-like grandeur, and off in the distance you can't help but notice the gleaming stars scattered in every direction.

You scroll your mouse forward and travel across space at the speed of light, coming across the Milky Way galaxy. Zoom in further and you will approach the familiarity of our solar system—the planets obediently and elegantly following their path through spacetime as if programmed to do so. You are manipulated by this never-ending sheet of spacetime with the world of the big distracting you from that of the small. One will stretch their mind so much to incorporate this grand nature that they will lose perspective to the truth that lies right in front of them: within them.

Yet you see a comet located kilometres from your home planet and you zoom deep into its composition. On your way through, molecules of water and carbon dioxide cover your monitor. Further in, individual atoms of hydrogen, carbon, and oxygen appear.

Congratulations, you have reached the subatomic world: a 'universe' of its own. The empty space within this atom seems overwhelming and infinite, yet in the distance you catch a glimpse of what we know to be an electron. Gleaming in the centre of this miniature universe you spot the hub of this network, the nucleus. Your curiosity taking over, your mouse sweeps across the screen and zooms closer to this entity.

As you are near, you notice that what appears to exist as its own is in fact comprised of many more individual entities, neutrons, and protons. You choose the proton and continue your journey further, reaching individual entities known as 'up' and 'down' quarks. You zoom in more to examine the foundation of these quarks; however, your magnifying glass appears to be unresponsive. It seems that you cannot dive any further and you are curious as to why. Well, every mind in control of this metaphorical magnifying glass questions the exact same thing. However, at this point our knowledge hits a wall.

According to our current understanding of the subatomic world, quarks are the smallest building blocks of matter. They are the absolute fundamental entity of existence. But just because our species may not yet have the ability to prove a smaller entity does not mean it doesn't exist. As humans, it is within our capability to observe reality and truth, not create it.

Because of this, we really don't know if our magnifying glass stops at the quark, contrary to how it may seem at this point of man's scientific journey. Although we haven't been able to prove it, the controversial concept of "string theory" attempts to pose explanations regarding the smallest entity of existence.

My knowledge of this theory is quite limited, but to put it briefly, this idea describes particles to be comprised of one-dimensional vibrating strings of energy. Along with this, the mathematics behind this theory require many 'dimensions' leading us to the question of why we don't experience such. String theory responds quite theoretical, suggesting that these extra dimensions are folded up in such a dramatic miniscule value that they prove to be irrelevant to us. Once again, this theory is just one interpretation of the universe striving to reach the prestigious bridge of truth.

I was in Grade 9 when I first created my own theory that I am about to explain; however, as my understanding of the universe exceeded that of a curious fourteen-year-old, I realized that my concepts were more applicable toward the philosophical side of existence rather than the physical. I later understood that what I was trying to say was still applicable to the world of physics but was just missing some sort of specific physical entity to say the least. Hence, I adopted the title of "strings" to describe what I was trying to say in a manner less vague. This will be explored later on.

My original motivation in creating this theory was to attempt to describe what I thought the concept of 'time' entailed. It later grew into posing potential theories for the other major questions I had. The more I dove into my ideas, the more I realized that all of the questions I listed above related to each other. So, if it seems like

I am jumping back and forth between concepts, it's for a purpose. It will only make sense if I go back to the concept of strings after I take the time to go through my progression of ideas.

GOD AND THE PRINCIPLE OF SCIENCE

> *"It is by the path of love, which is charity, that God draws near to man, and man to God. But where charity is not found, God cannot dwell. If then, we possess charity, we possess God, for 'God is Charity.'"* (1 John 4:8)
>
> —*Saint Albert the Great, Patron Saint of Natural Science*

When I ask myself "How does God guide me?" I think about prayer; I think about love, the Bible, or my "Alpha" group; and I also think about 'science,' specifically scientific inquiry. By design, our species is always drawn to asking 'why.' We ask why things are the way they are. This is in our design because it is part of God's plan for each of us. We ask these 'why' questions because we care about discovering that which is deeper than us. Specifically, it is the love and the passion that allow us to care—the love and passion that God gave to us at the birth of creation.

When specifically looking at the scientific method, the prime focus is to question why things are the way they are. We ask why nature behaves the way it does. Almost like a puzzle, we search for these pieces that fit together, these theories that complement each other. However, the more we search and the deeper we go, the more we hit a realization that there are parts of nature we simply cannot explain. The deeper we go on our scientific journey, the more we hit a realization that there is beauty beyond physical phenomena. There is a beautiful reality that cannot be explained by any physical law.

I often wonder why it is that our species cares so incredibly much about that of which we do not know or understand. Why do we ask these 'why' questions all so frequently? Science, after all, is built on the foundation of these types of questions. All our discoveries, our technology, and our medicine are results of us asking these questions. We constantly strive to maximize our potential as a species, and it all starts with asking both 'why' and 'how.' Curiosity, commitment, and passion are all born out of love, and God is the only true giver of this.

> *The First Letter of John 4:7-13 – Beloved, let us love one another; for love is of God, and he who loves is born of God and knows God. He who does not love does not know God; for God is love. In this the love of God was made manifest among us, that God sent his only begotten Son into the world, so that we might live through him. In this is love, not that we loved God but that he loved us and sent his Son to be the expiation for our sins. Beloved, if God so loved us, we also ought to love one another. No man has ever seen God; if we love one another God abides in and his love is perfected in us.*

Yes, not one man but Jesus himself has truly 'seen' God, but like this passage notes, God abides in us and he only abides in us through love. God abides in us through love for one another, through love for the nature around us, and through love for the beauty which is deeper than us.

This is why we, human beings, are the way we are—it is why we are drawn to science and hence drawn closer to God. The most fundamental and essential component of a true scientist is *love*, because this leads us and our minds to places much deeper than our observable surroundings. This love was the fuel behind all my ideas as a young physicist, many of which you will start to learn.

MULTIDIMENSIONAL EFFICIENCY

Before I start this section, it is valuable to point out that what I am about to write was one of my first philosophical approaches to making sense of the physical world. Although I have elaborated and changed a lot of what I am about to describe, these ideas were the stepping stone for my current theories regarding these 'dark waters' of our universe. During the next few pages, I talk a lot about 'energy.' I use this term quite ambiguously. For now, take it as it is; I assure you that in a few sections my explanation of energy at the quantum level will tie all of these ideas together.

My interest in the concept of energy started in Grade 9, shortly after I was first introduced to the law of conservation of energy in my science class. I learned that energy cannot be created nor destroyed, yet I was slightly confused with regard to what this entailed. I realized that the actions performed within our universe involved the conversion of energy from one form to another; hence, I understood that no 'task' was ever one hundred per cent efficient because of this underlying rule.

However, my knowledge of this opened up some doors of curiosity within my mind. I started thinking about the universe at a larger scale, specifically its expansion, and I wondered if the same principle of energy applied.

I wondered if this law ensured that the expansion of the universe would be prevented from being one hundred per cent efficient.

Regardless of how limited my knowledge was at this point, it made sense to say that our universe was drenched in energy. It must have some sort of bank of energy—right? Well, I settled with yes, the universe did have a 'supply of energy' which it could use to rapidly expand ever since it was created. However, if it were to follow this underlying law of physics, its task of expansion would not be entirely efficient. If not *all* of the energy required for expansion was actually used for 'expansion,' then where does this lost energy go? In other words, what was this transfer of energy?

To put it simply, my initial idea had been that this energy was transferred to a new *coordinate of dimensions.* The concept of a 'dimension' has a very broad connotation, and I am fully aware that science fiction has created this concept to encompass an incredibly 'fictional' footprint. I am not going to go into detail about the logistics behind varying coordinates of dimensions in this section, but for now I just want to be clear about the connection between *dimensions* and the entity of a *universe.* So, for the remainder of this explanation understand that differing combinations of dimensional coordinates result in various possible entities—that of which we call a universe.

It is important to understand that our current universe has three spatial dimensions, regardless of how technical you can be. I believed that there are in fact more dimensions, or at least the potential for more dimensions that haven't been activated. I thought that maybe the unleashing of these dimensions would happen in future universes. Hence, the only way they would be activated is if they had enough energy to 'set them free.'

Maybe only when an energy threshold is met will there be a creation of another universe encompassing not the first three spatial dimensions, but possibly spatial dimensions four, five, and six. To sum it up, this future universe currently exists as nothing but potential until enough energy escapes our current universe.

But how is this energy threshold met? Because no event in the world of physics is ever entirely efficient, some of the energy used to expand our universe will be lost—the conversion being the transfer of energy to this future universe. Only when enough energy is available to spur these other dimensions will it have the ability to become 'unleashed.'

Now, this idea led to my understanding of *'true time.'* If the energy from our universe was gradually, yet constantly, becoming transferred to a future set of dimensions, there will have to be a time when there won't be enough energy to sustain our universe. At this point, all of the energy that spurs our universe to life will escape its boundaries and lead to the birth of a new universe. In other words, this will be the 'end of time' for *us*.

True time has nothing to do with distance, nor the speed of individual entities within space—the universe could care less about the movement of its occupants. True time is a measure of accessible energy. Right from the start of my scientific journey I've made a conscious effort to connect the concept of time to the 'energy of the universe.' For now, what we define energy to be is irrelevant. What we care about is the ubiquitous correlation between time and energy.

THE "LATERAL TIME THEORY"

As my ideas grew, my concept of the 'energy reservoir' became immensely more specific; however, at this point it is enough to say that this reservoir of energy simply refers to the total energy accessible to the universe. To describe this, I created a simple concept named the *lateral time theory*—lateral meaning extending from the sides (referring to how I first envisioned the concept).

Imagine a number line extending indefinitely from left to right. Pick a point and assign a value to it. I will call this point '1000' to make it simple. This number is the total energy of not the universe, but all of 'existence,' and this point is not a universe, but the creation of such. In our case, this is the Big Bang.

Going back to my 'cause and effect' principle, this initial Bang would be referred to as the cause. Hence, it creates our universe, or in other words the transition. This can be imagined to be the space in between the points of the number line. Once again, in my eyes the reason why something starts is to achieve its end, and the only way it does so is through its transition.

In terms of the lateral time theory, the purpose of this transition is to transfer the energy of existence from our universe into the expression of potential dimensions. This is the order of the universe: the task it was created to do. The majority of energy is used for this transfer because this is the ultimate *order* of the universe. In 'human terms,' when we lose energy through energy transfers in our day-to-day tasks, it is quite minimal—we are efficient beings.

This contrasts with expansion, however, as its sole purpose is to 'lack efficiency.' If it *were* entirely efficient with its 'energy,' the 'other dimensions' would not be activated. We will only be able to draw the next point on the number line as soon as this pristine order is completed. This new point is a new cause—a new Big Bang to say the least. However, instead of our original energy value of '1000,' the completion of the previous expansion may result in a current Big Bang energy value of '999.99' for example.

In case I haven't been clear, these numbers are not meant to be taken literally. Rather, they are to be understood as a comparison to the ratio of accessible energy built into the root of our universe. Yes, I understand this may not be feasible. If the universe used up energy during expansion, what field is it expanding in, and why would something, like energy, that exists just disappear?

Surely it's quite simpleminded to say that just because the past is a reference to our universe at previous points of expansion, this energy is entirely gone from existence. However, keep in mind that I didn't question this at this point—that comes later. After all, this book is simply a progression of my mind's wandering throughout my exploration of our rather questionable universe.

We have now established that our universe will eventually express the potential for another 'cause';

however, it doesn't stop there. To make it easy, let's say that the Big Bang was the first cause, or the first point on the number line. Our next cause would then be our second point. It is now in the exact same position our universe had been at its creation. Hence, it now has the potential to continue on fulfilling the order it was given by the previous universe.

So, the second Bang (the moment all the energy needed to express specific potential dimensions were given by the preceding expansion) will once again spur an expanding universe. What makes this Bang unique from its parent universe is that it is given less energy to fulfill its order. This 'cause' was created to reach its 'effect,' and this requires a 'transition.' This is yet again another expanding universe incorporating a new set of dimensions.

This sequence will continue to occur across our metaphorical number line, with each cause driven to fulfill its order. However, it will do so with less accessible energy than the last. Now, would this number line come to an end at some point? If this energy value of '1000' is ever so steadily decreasing, wouldn't there come a point when there simply won't be any energy to sustain a 'cause'? Our universe's piece in the puzzle of true time will end as soon as our order is fulfilled; however, *true time* itself will never end.

Every cause created from its parent transition will use a portion of the transferred energy, but if you define a portion in a mathematical sense, this is referring to division rather than subtraction. No matter how many times the original value divides itself up, there will always be a value. However small this value is, it will still exist.

Yet there will become a point when the available energy in the energy reservoir becomes so incredibly low that it will become extremely weak and ineffective to say the least. In this case, expansion will *slow down*.

As you can imagine by now, this affects how much energy will be transferred to expressing the potential of future dimensions. Without meeting the threshold of energy, these dimensions will not become activated. This extremely weak expansion is what I call "*Quantum Freezing.*" The transition will 'freeze,' to say the least. This universe will still exist, yet without a deliberate expansion, the activation of future dimensions will be incredibly delayed.

There *will* be a time when a universe becomes stuck in a permanent state of transition—enough energy will be left to sustain its existence, however, there will not be enough energy to fulfill its 'order.' This state will exist indefinitely—it will be a universe that denies change. This universe will be 'stuck in the present,' you could say. However, our accelerated expansion shows that our place on the timeline of existence may be far from reaching this point.

Before I wrap up this section, I will admit that it's difficult trying to explain what went on in my brain years ago. We really are biased by the present. So much of what I have just written I've changed and developed. There are many elements of our universe I have yet to cover and this excerpt is by no means a way of ignoring the inevitable.

Regardless, this is a crucial philosophical and physical foundation for future sections in this book. My ideas here built the foundation for my explanation of the pretentious concept of time, the "purpose" for the creation of the universe, the need for expansion, and the questionable idea of a depleting energy reservoir—those of which I will most definitely come back to.

TIME TRAVEL

Now, let's start this section by talking about "time travel." As I flip open my journal (quite a few pages back), it appears that the next thing I scribbled down was my idea regarding this concept. The ability to travel backwards through time may sound incredibly fictional; however, when you really think about it—it's a genuine idea. What is it that makes the future different from the past?

Well, let's start with entropy—a measure of disorder. To put it simply, our universe's past was incredibly ordered. In its early days, everything encompassing our cosmos was unified in every way. This presence of unification was so strong that it continues to exist today. You will be expanding at the same rate regardless of where you are on the map of the universe.

However, though it's still held together by this original element of complete unification, expansion is slowly deteriorating this bond of order. In every moment since you've started reading this book, our universe becomes more and more disordered. Perhaps this is time's 'referee.'

Keeping this in mind, go back to assuming that energy is constantly diminishing throughout the expansion of the universe. In this case, available energy would distinguish the past from the future. Hypothetically speaking, if you were to go "back in time" (now this is my concept of 'true time' we are speaking about), you would be going to a state encompassing a greater amount of energy. This is completely nonsensical. If your time machine were truly able to travel to the past, you would be *reversing expansion*.

Now, even this is incorrect. Really, a reverse expansion in the context of energy will be exactly the same as a forward expansion. The overall movement of the universe (whether it is expanding or contracting) is still draining its energy reservoir. Regardless of direction, energy is needed to fulfill this "connection" to an end result. It is important to realize that distance is a property of our universe—and only our universe. In whatever 'arena' this expansion is taking place, the concept of distance breaks down.

I go into the whole concept of existence outside of our universe in a further section, but for now, just understand that the beginning of the universe and the end of the universe are *NOT* separated by distance. So, what depletes this energy reservoir is not a factor of travelling across a distance, but rather the formation of a connection between what *is* and what *could be*.

If an entity is in a current state and it is drawn by a connection to another state, the bond that develops this connection requires energy. Once again, the concept of energy is quite arbitrary, but I will save a proper explanation of the nature of energy for later. Whatever phase in which our universe currently exists, *what it is* differs from *what it could be*—this being the future.

If something is in a current state of being, it is everything but what it's not. In order for this state to get to 'what it isn't,' there must be a bond of connection between such. This bond is this language of energy. If you have a fixed view of what you envision the concept of energy to be, it is best to read this section with a fresh mind.

Let's go back to 'reversing expansion.' If you were able to contract the universe, you would still be "travelling forward in time." Energy is required for an entity to develop a connection to what something could be (in this case the universe is not contracted on itself, yet it would be with "reverse expansion")—and the process of this will drain the energy reservoir.

Regardless of what something does—if it does anything at all—energy will be used. This is what causes the constant forward movement of true time. There is no possible way that one could 'gain energy back' and truly travel to the past.

Now, let's talk about travelling into the future. From the framework of cosmic inflation, we are travelling forward in time every moment. We are going forward in time because of this 'loss of energy.' During every individual moment of expansion, the energy reservoir is draining. However, instead of draining all at once, the process is ordered, even, and gradual. To put it simply, energy cannot 'skip values,' and this is due to the 'order of entropy.'

Think of it this way: if you were to draw a square a certain size but wanted to make it smaller, you would minimize it. (Assume you are able to minimize entities on paper as if you were on a computer.) This shrinking process isn't instantaneous. Rather, in order to get to your preferred size, you are required to pass through a series of bigger sizes, and this is called the transition phase.

If there were no transition phase, the minimized square would fail to exist as the same entity as the original square. This goes for anything. For something to remain as it currently exists, it must have a connection to what it *could be.* This connection is the transition, and this transition phase is what allows an entity to remain in harmony with the product of its change. Without this harmony, you would be looking at two completely different squares, each encompassing their own existence.

But what does this have to do with travelling into the future? If we were to travel to the future, we would simply be skipping a crucial transition phase of our universe. For now, let's say we are skipping the process of energy usage. In this case we would have two entities, each occupying their own existence. By 'travelling to the future' in a true context, we are breaking the transition phase of our universe, hence breaking the harmony of unification.

Because of this, the future is no longer the future but a separate entity of existence. So, it is crucial that our universe follows the order of its transition. Take away this transition even the slightest and our universe will cease to exist as it only does so through order and unification. Hence, the expansion of our universe must happen fluently. The energy decrease happens in a pattern of order, and only with this order will our universe sustain itself.

I completely understand if you finish this section and some terms may seem a tad ambiguous. I assure you that the concepts of energy and order are thoroughly explained during our exploration of the universe at its most fundamental level. With this section, all I hope for is that you are able to grasp the impossibilities of time travel within the concept of my little theory of *true time*.

'DEBRIEFING' TIME

I do apologize if the last few sections were *slightly* overwhelming, or perhaps difficult to grasp. I wrote these when I was a little younger than I am now, but these were little 'theories' I was quite proud of. They contributed to this puzzle my mind had always tried to stitch together, but like the other sections of this book, I would like to integrate a faith-based approach to understanding the concept of time. To begin, I would first like to include one of my favourite passages from the Holy Bible—one out of Genesis.

> *Genesis 1:1-2 - In the beginning God created the heavens and the earth. The earth was without form and void, and darkness was upon the face of the deep; and the Spirit of God was moving over the face of the waters.*

Sometimes I wonder if I think too far into things. In fact, I know I do, but sometimes I do think it is needed, or more importantly valued. The Bible is the word of God as he speaks to *us*. It is not a 'stand-alone' book. It is a book that needs to be read by every unique being, and it only plants its beauty when it is taken into the human soul. The messages that God relays through His word may speak to different people in different ways—it is not always up to us.

I always used to feel 'guilty' for making analogies to things or comparing texts to phenomena much deeper than my peers had, but I realized something. I realized that my brain was designed this way for a reason. If I wasn't—or if anyone wasn't—supposed to see things in a certain light—we wouldn't. Simply the fact that we as humans see not only the Bible but everything in our own unique light just reminds me that God has a different plan for each of us.

We are not meant to force our minds to walk in the path of our neighbour simply because we see this perhaps 'unique' light. We are only supposed to place our trust in God and continue our unique journey on the path He designed for each of us.

With regard to the passage from Genesis above, something unique stands out to me—something that for a scientist feels special: *"In the beginning God created the heavens and the earth. The earth was without form and void, and darkness was upon the face of the deep; and the Spirit of God was moving over the face of the waters."* I will begin with this: God did create our waters, our earth, and our heavens, and between them all He created our 'universe.'

However, when I think of the 'waters' mentioned in this passage, I begin to think of the grandeur of 'space' or cosmos—in other words the all-existing 'universe.' 'Darkness was upon the face of the deep'—the darkness was upon the face of the deepest waters imaginable. This is an infinite ocean, deep beyond understanding.

This ocean consists of the 'waters' of our cosmos, the gentle and peaceful flow of our cosmos guided by the spirit of our Lord. He is everywhere and in everything, and His beauty, His spirit, was moving these waters. He is these waters. He exists among this 'ocean' above us and is guiding the flow of water to His heavens above.

In the sections above I talk about the expansion of the universe, the constant and everlasting expanding cosmos above, but when I read this passage in Genesis, that is exactly what I thought about. God is not only guiding this 'expansion'; he is this expansion—this contestant movement of our grand universe to the beauty of heaven.

God is this movement of time, this constant movement of time into a beautiful future that is reached at the furthest depths of our universe and in the heavens above. When I think of the concept of 'time,' this is what my mind is drawn to. I think of the gentle flow of our beautiful cosmos to the grand heavens above. God is this creator of our waters, and He is the only one who can move these waters—who can move time. He is outside of time itself, but He is also time Himself. To me, this is absolutely beautiful.

> *Genesis 1:6-9 – And God said, "Let there be a firmament in the midst of the waters, and let it separate the waters from the waters." And God made the firmament and separated the waters which were under the firmament from the waters which were above the firmament. And it was so. And God called the firmament Heaven. And there was evening and there was morning, a second day.*

Like I have explored, God truly is the only one who can control the ever-flowing beauty of 'time.' I want to briefly relate this to my above discussion of 'time travel.'

In other words, the backwards or forwards 'leap across the ocean.' Like I explained using 'scientific' reasoning, it is not in our reality to 'leap across this ocean'—to jump across time whether that be to the past or the future. It is a construct: a playful idea founded by man.

> *Genesis 1:14 – And God said, "Let there be lights in the firmament of the heavens to separate the day from the night; and let them be for signs and for seasons and for days and years."*

'Time' was thoroughly designed by God to flow one way—and that was to heaven, the future that is our final destination. This light of God in the firmament of heaven is guiding us and the waters of our cosmos to the final destination: heaven. He is in control of these 'seasons'—the different milestones in our earthly lives. God's unique plan for each of us is shown through His love for us, His light. It is a path we follow, and it is the unfolding of certain events in our lives that leads us to Him. It is our duty to follow this path, to live every day and grow—to age and to change.

'Time' is not something that is in our control; it is God's waters we are swimming in—we can't change that. We cannot skip to our future, skip the 'season of winter' to enjoy the beauty of 'summer.' We must flow through all of the seasons, all of our life's stages—one at a time. We also cannot 'jump back' to our past; what's done is done. This always reminds me why God blesses us with His gift of forgiveness. The flow of time is something that we all experience; it is in us, but it is not something we have control over—that is God's job.

> *Acts 1:7 – He said to them, It is not for you to know times or seasons which the Father has fixed by his own authority.*

There are some passages I would also like to include. They are little, almost subtle reminders that time is in God's hands and God's hands only. Both of these are from Ecclesiastes, a chapter of the Bible I am quite fond of.

> *Ecclesiastes 3:1-9 – For everything there is a season, and a time for every matter under heaven: a time to be born, and a time to die; a time to plant, and a time to pluck up what is planted; a time to kill, and a time to heal; a time to break down, and a time to build up; a time to weep, and a time to laugh; a time to mourn, and a time to dance; a time to cast away stones, and a time to gather stones together; a time to embrace, and a time to refrain from embracing; a time to seek, and a time to lose; a time to keep, and a time to cast away, a time to tear, and a time to sew; a time to keep silence, and a time to speak; a time to love, and a time to hate; a time for war, and a time for peace. What gain has the worker from his toil?*

> *Ecclesiastes 3:11-16 – He has made everything beautiful in its time; also he has put eternity into man's mind, yet so that he cannot find out what God has done from the beginning to the end. I know that there is nothing better for them than to be happy and enjoy themselves as long as they live; also that it is God's gift to man that everyone should eat and drink and take pleasure in all his toil. I know that whatever God does endures for ever; nothing can be added to it, nor anything taken from it; God has*

> *made it so, in order that men should fear before him. That which is, already has been; that which is to be, already has been; and God seeks what has been driven away.*

Before I go on to the next section, I want to connect a little bit of what I just wrote to my own life. I always used to wish there was a 'rewind' button to life. I always wished, I wished so hard, that I could go back in time and change certain events in my past. I think this is a reality quite a few of us have wished for. Unfortunately, this is not reality.

I always used to think about how much I could change my life for the better if I could go back to my past and guide this little girl who had lived in pain for so much of her life. I wished that I could go back in time and give 'little me' the wisdom to seek help for my struggles—to not take my frustration our on myself and to not blame myself for an illness that was anything but my fault. I so, so wished I could do this.

However, the older I got, and the more wisdom my experiences granted me, the more I realized that I wouldn't change a thing. I would not change absolutely anything about my past. The obstacles, pain, and adversity I faced as such a young girl gave me the gifts I would not have gotten any other way. They gave me perseverance, they gave me courage, and they gave me strength.

But most importantly, they gave me compassion. They gave me compassion for those children who had to go through what I had, and my struggles gave me such a passion to help those. We are all God's little lambs. Some of us may get lost a little easier, or some of us may be a little more crippled than the rest, but our struggles flourish into our greatest strength. Ask anyone with a story like mine, and they will tell you the same thing.

I didn't know it at the time, but my struggles led me to a place in my life I would want nowhere else to be. I wake up in peace, and I start my day inspired. I wake up inspired to share my story with the world and inspired to help others who may be struggling like I was. Once again, this is why I am writing this book. I am writing this book not only to showcase my creativity of philosophical 'theories' of our universe, but to inspire others that there is light at the end of the tunnel—such a beautiful, powerful light.

ENERGY AND THE QUANTUM—PART 1

Now that we've seen how our universe behaves at the macroscopic scale, or rather a philosophical approach of such, it is only suitable to talk about its existence at the smallest of scales—a scale beyond observation. It is incredibly easy to become caught up in the grand scale of our universe, developing theories only to satisfy its large astronomical behaviour.

However, one fails to recognize that this large entity we call our universe is simply just a reflection of the small. The laws of physics, the pristine order, and the systematic progression of the universe are all just a reflection of quantum scale processes. These processes are beautifully inflated to tell the story of the unobservable.

It is naive to believe that the astronomical behaviour of our universe functions at random. For something to be considered "random," it must be compared to a state of order. However, this "order" is yet to be established. Without such as a frame of reference, there would virtually be no difference between systems of chaos or order.

Hence, origin is not a product of randomization, but sub-measurable order. Something allows for such an ordered, collected system to function, and this can only be explained through an analysis of the most fundamental ingredients of our universe.

'The most fundamental ingredients of our universe'... Perhaps our 'cosmic magnifying glass' really does hit a wall at our electrons, or our quarks. Perhaps this is the only provable hypothesis, and perhaps it is the logical conclusion to settle on. But perhaps we can extend our scale a little further. Really, there should be a point where the smallest of scales has the ability to explain the largest with such precision and accuracy so that quantum to cosmic unification can be confidently expressed.

However, if we stop this scale at its current position, I am not confident this will ever become a possibility. What we are missing is cosmic communication amongst the grandest of scales. After all, this large entity we call our universe is just a reflection of the small. Now, take my words with a grain of salt as I attempt to explain my thought process... my wandering daydreams and my imagination.

When I started thinking about scales beyond the empirically observable, the first thing that came to mind was the concept of 'string theory.' Now, quite frankly, I have absolutely no fluency in the dialogue of this theory. I remember hearing it referenced in many popular science books, but other than its name I am not very sure what it preaches. However, simply the concept of a 'string' is what got the wheels turning in my imagination.

Regardless of the 'shape' of this subatomic entity, I envisioned it to appear as an infinitesimally thin strand, a 'string-like' entity capable of both rotation and vibration in the most fundamental sense. If we go back to one of the beginning sections of this memoir, I mention the concept of 'true time.'

I explain this through the 'lateral time theory' in which the 'purpose' of universal expansion is to 'unleash' further dimensions. I go into a generalized explanation in saying that as our universe expands, it uses 'energy,' and the more energy that is lost in the specific set of dimensions, the more 'potential' a future set of dimensions has to become expressed, or in other words 'activated,' leading to a series of 'cyclic universes.'

Now, this is purely a philosophical thought. I throw around the term 'energy' quite loosely, but when I talk about this theory at a smaller scale, I hope to bring a bit more technicality to the concept. In fact, when I started thinking more about strings, my goal became to create a rational explanation of universal expansion. I pretended that the fabric of our universe consisted of individual 'string-like' entities that occupied every direction and spatial dimension.

However, I imagined that all of these strings were 'laid out' one by one. They were not connected or overlapping in any way, but what each 'string' had was its own 'field' which I called a fundamental field or 'string field.' Each string radiated this field, and the fabric of the universe was 'born 'through the strength of these fields, i.e., a unified universal field to say the least. (To go on questioning *what* these strings and fields consist of is purely a philosophical inquiry.)

The point of this thought experiment is not to question deeper than this utmost fundamental entity, but to offer an explanation to their predicted behaviour. It is important to note, however, that these 'fields' shielding each string are not a 'physical' entity, and neither are the strings themselves—this is explained later.

Once again, I concluded that these strings are infinitesimally thin entities, and that their fields make up the 'fabric of space.' More specifically, these strings both vibrate and rotate in relation to one another, but it is the *rotation* of these strings that causes 'expansion.' Perhaps it is the orientation of these subatomic strings that differentiate the sets of dimensions I mentioned in the lateral time theory.

As the string rotates, it is changing its 'orientation'; however, if we were to denote string orientation to a dimensional coordinate, the more this string 'rotates' away from its initial position, the further the universe is in its 'transition state.' Hence, the further our universe is along its 'transition state,' the higher the entropy becomes from the reference of our current 'dimension.' In other words, our current spatial dimensions 'fall apart' the more the string rotates, as the strings are consistently *transitioning* to the 'next set of dimensions' (which is spatially incomprehensible to our current universe).

Referring back to entropy, this can be considered more of a subjective concept than anything. Yes, the total entropy is increasing through 'expansion'—our universe is becoming more 'disordered' the more these subatomic strings rotate. However, this is only from the perspective of the initial position of these strings, i.e., the 'beginning of our universe.'

The more something 'changes' the less unified it is with its initial state, but at the same time the more this entity changes, the more it becomes one with that which it is becoming. So, the entropy from the perspective of the 'final position' of these strings per se is actually *decreasing*. According to this 'future moment' in which the strings rotate to their 'new' orientation, the universe will have been immensely ordered.

ENERGY AND THE QUANTUM—PART 2

Now, previously I had gone on to talk about the process of energy transfer through universal expansion, and I would like to elaborate on that a bit. In my explanation of the 'lateral time theory,' I had described how as the universe expands through its inflation, it is draining its 'reservoir of energy.' That lost energy is then 'transferred' to a future universe, which in turn encompasses a new coordinate of dimensions.

I envisioned this concept long before I started thinking about quantum strings; however, I realized that in a way it applied. I mentioned above a type of energy field surrounding these subatomic strings, and if we were to pretend this represented the 'energy reservoir' of our universe, I could perhaps bring some physical sense to the philosophical nature of my 'lateral time theory.'

To explain this, I would start by bringing attention back to the rotation of our subatomic strings. To summarize, the further along they are rotating, the closer they are to reaching a specific orientation. Only when they reach this orientation will a new set of spatial dimensions become 'activated.' When this point is reached the spatial dimensions of our current universe will have gradually 'fallen out of place,' but in their replacement there would be a highly ordered set of newly activated dimensional coordinates.

These dimensions would then have very high entropy and would continue to rotate until a new 'orientation' is reached, and only once this correct orientation is reached would another 'new universe' be born. This cyclic progression will continue to happen, creating a new universe every time a specific and unique orientation of strings is reached.

However, as I mentioned at the beginning of this section, these strings are each surrounded by their own field. As our strings rotate, this field no longer becomes a field for orientation A, but for orientation B, and then later for orientation C. The 'energy' that was surrounding the initial orientation of the string now becomes the energy surrounding a newly orientated string. Keep in mind that this is still the same string; the only thing that is changing is its 'rotation.'

But, once again, as the string rotates and reaches its specific orientation (one that gives rise to a new set of 'dimensions'), the energy field now becomes a field for these new dimensions. At every single moment of rotation, the strings are one step closer to being in a highly ordered state with potential to unleash a new 'universe' with vastly different dimensional properties. Hence, the field that encompasses these strings will encompass every stage of rotation. And once again, at every stage of rotation, the strings are making their way to a very specific orientation—an *'unborn' coordinate of dimensions.*

Therefore, the 'energy of our universe' is constantly progressing toward becoming the energy behind a future universe. Like in the lateral time theory, this entity of energy *is* making its way to a new set of 'dimensions' (orientations) as the spatial fabric 'expands' (string rotation). So, yes, the energy reservoir of our current universe is 'draining.'

At every moment of expansion, energy is in a way *transferred* to the next 'set of dimensions,' and this is a process that will go on indefinitely. Each universe will be unique in that it will be dimensionally distinct from its parents, and this state will be vastly different from the nature of our own three spatial dimensions. Yet each universe will still begin with a Bang and still continue this progression of string rotation.

THE CREATION STORY

We have learned about the 'field' behind the fabric of our cosmos, but what about physical entities such as matter or light? In this chapter you will learn how my explanation of string cosmology ties into this.

This is another section of my original book I was hesitant to include. It is purely a 'game of thought' and is not scientific in the slightest. It is also quite specific, but it is purely my mind playing with little ideas. This section does tie into the one before it—the little explanation of cosmic inflation and quantum strings—but it hopes to shed more light into the behaviour of these strings.

Once again, I am quite embarrassed to be writing about such fundamental entities in such an 'un-scientific' way, but I guess maybe this was just my mind trying to take a more creative, 'puzzle-doing' type of approach to answer the deep questions that kept me up at night. I always wondered what was beneath the depths of our most fundamental particles such as an electron or a quark. Hence, I 'developed' the little theories regarding the flow of 'quantum strings.'

But I also wondered this: at the very most fundamental level, what constituted the differences between particles of 'mass' or massless particles such as photons? I also wondered what comprised the fabric of our cosmos, a completely non-physical entity, but surely an entity of 'space' nonetheless.

Well, let's start with a physical particle—an electron, for simplicity's sake. I begin by pointing out that beneath the blankets of the electron, there may also be these 'quantum strings'—just like the passage about the flow of 'space' I wrote above. Except, as space itself is different than a physical entity, these little 'strings' may behave differently than I mentioned above. (I do realize that an electron is not an entirely physical entity, but for this section, the little girl who created these ideas years ago pretended it was.)

I remember thinking to myself, hmm—the fabric of our universe, space itself, can perhaps be thought of as a series of these independent and non-adjoining strings. And what I meant by 'non-adjoining' is that they were each separated by their own 'energy field', free from contact with one another.

I'm going be a little playful and ask you to imagine a baby worm floating around in its own little bubble through space, doing its little dance of 'rotation' through a series of dimensions.

Quite silly, I know, but this was pretty much what I originally thought. Once again, space, I thought, was a non-adjoining string-like entity within its own energy field. I called this field an 'initial field.' It does not exist as a physical entity, as on its own this fundamental 'string' ceases to be a physical entity. However, if more than one of these little 'strings' join together, there is an ability for 'substance' to form, with substance being a massless or a massive particle.

This electron does exist as a wave, but in a different light I chose to look at it as a rather 'unpredictable' massive particle. (For those not well versed in scientific dialogue, massive does not refer to a 'large entity' but rather one that simply has mass.)

However, the electron's fundamental 'strings' are connected, unlike those of spacetime. Maybe it is comprised of only two strings, or maybe three, or maybe one hundred—that doesn't matter for now. What does matter is that these strings are connected and overlapping, and their own 'fundamental string-fields' are merged together—in union almost as if 'two become one.'

As a little girl, I wondered if perhaps the union of these basic, dimensionless, and massless strings becomes an entity of 'substance.' I wondered if something about the union of the most fundamental field imaginable created the building blocks of our everyday lives. This is what my mind dreamt of years and years ago.

Now, if we were to look at a little different situation—the behaviour of 'strings' which have the potential to create a 'massless substance' such as a photon—things would look a little different. Like the behaviour of strings beneath a substance of massive nature, the strings of a photon would also be connected or 'attached.' However, their field—the individual fundamental of each string—would not be in union with each other.

Once again, imagine two little worms connected at the tail. Now imagine that these little worms had a 'halo' encircling their head—the area of the strings at the ends farthest away from one another. This 'halo' is the individual strings' little field of energy, or the fundamental string field referred to above. These strings were in fact connected, and this is what gives a photon substance—the substance known as light.

However, the fields of these two strings were *not* connected, or in other words they were not in *union* with each other. They still existed as their own unique and distinct entities. Years ago, this is how I distinguished the physical from the non-physical, or the non-substantial entities from those of substance. To me, it was the intimate union of these powerful fields that had the potential to create 'substance.'

When I thought of a photon, I didn't think of it as a little string encompassed by its 'bubble' of an energy field. Rather, I once again thought of it as a connection of more than one string in which their little energy fields, their little halos, are only at the ends or the 'tips' of the strings. The energy fields were then separated by what is known as its 'string.' I thought this was important to point out as my younger brain began to wonder why light almost 'flows' through space.

To me, light just seemed too elegant in the way it drifts across our cosmos. It doesn't need a direction—it just reflects out into the deep waters of space. It always seemed like 'light' had some sort of 'union' with space. I then began to think—what if it did in fact have a 'union' with this entity of spacetime? The fabric of our space was once again what I thought to be a collection of strings independently encircled by their own fields. It was unlike a photon whereby the strings were only partially 'haloed' by their fields; rather, it was entirely at one with its field. A diagram of these concepts is included.

A: Representation of the basic string and singular field of 'spacetime'

B: Representation of joined stings and the union of fields- the components of a 'massive' particle. For example, an electron

C: Representation of joined strings but separated fields- the components of a 'massless' particle. For example, a 'photon'

I thought that light flows so eloquently and independently through space because a portion of each string 'merged' with the field of space itself—specifically, the portion that lacked the energy field. Remember, I believed that only a little portion of the string was encircled by its field; hence, the union of the strings that weren't entirely covered by their fields produced what can be called a 'photon.' The rest of the string was essentially 'unclothed.' It was separate from its field. However, that part of the string would have needed a 'medium' to travel in—or else perhaps light wouldn't travel at all? This is what I thought at the time.

So, I imagined that light itself, the unclothed, bare element of the 'string,' became in union with space. It travelled quite literally though space—through the fabric of energy encompassing every and all areas of our cosmos. Light was the entity which had the potential to become one with space, and so it did.

Another thing that always puzzled me was the fundamental 'difference' between varying subatomic particles. An electron, I thought, is simply an electron. However, a proton, or neutron comprised of quarks, puzzled me a little bit. Within the differing species of quarks, was there anything specific that differentiated them? Perhaps the vibrations of these subatomic strings? I do remember hearing something within that nature in 'string theory.'

So, let's say that is the case. However, I was a little worried about this—it seemed a little flawed. There are only a limited number of 'subatomic' particles, yet there can be an infinite number of vibrations? Why, then, would there be only a selection of these subatomic particles in our universe when there potentially could be an infinite number of them?

To explore this, my thought process returned to my chapters entitled "Energy and the Quantum." Part of the puzzle I constructed was that these strings of our cosmos rotated ever so slightly as time went on. Well, they were 'time' itself. But, as soon as these strings reached a 'specific' rotation, they reached a threshold of a new 'dimensional coordinate.' This specific orientation of these strings would in turn cause an event similar to what is known as the Big Bang.

I thought perhaps this 'infinite' vibration of our subatomic strings really does exist. However, maybe many of these vibrations cease to exist in our current universe— our current coordinate of dimensions—because they are simply 'incompatible' with the specific rotation our 'strings' are presently in. Maybe these vibrations will happen eventually, but only when our strings reach this very specific dimensional rotation. Maybe the uniqueness of the rotation could perhaps depend on this vibrational pattern.

To me, this made sense. Perhaps that is why we don't 'see' many of these subatomic particles that in their right have the potential to exist. Maybe this is just because they won't exist in our universe. Maybe they simply can't. In this case, each dimensional rotation of the fabric of strings known as our universe would heed very different particles. This idea once again led me to another piece of the puzzle—the theory of the Big Bang.

I have my own beliefs on this theory now, but I want to talk about what I believed when I was younger. I never really thought of the Big Bang is some humongous and catastrophic explosion; instead, I thought of it is something subtle yet still so incredibly powerful. At the very end of our own universe and at the very moment of the beginning of the next, there would be a very subtle 'switch' to a specific coordinate of dimensional rotation. Specifically, it would reach the threshold needed to 'spring out' a new universe unique to those specific 'dimensions.'

This subtle threshold was essentially the Big Bang, but the Bang was caused by something else. At this very moment of rotation, some of the vibrations—the vibrations which could only happen at a *specific orientation* of these strings—would 'unleash' themselves. It is at that very exact moment when a new 'set' of subatomic particles is born.

It would happen instantaneously at this very specific moment of rotation—the moment when there is a potential for particles to form. I say potential, as the *concrete* formation of these 'massive' particles would not happen instantly. Remember that a particle of mass would require the *union* of its energy fields. This process would *not* happen instantaneously.

The formation of these particles would start off as what would appear to be 'light'—a variation of a massless photon. At the moment of this Big Bang, the many strings of our cosmos would 'peak through their fields' and attach. This is when a new set of specific vibrations would occur.

However, the strings would attach at their 'tails,' like in the diagram of a massless particle above. Because they would attach at their 'tails,' the energy field of these strings would not yet be in union. They would form originally from what appears to be 'light'—for example, a photon as shown in the diagram. Although, as time progresses and the orientation of these strings change, the field of energy would be displaced from its original position and eventually be in union with its paired field. This would then be the concrete birth of a 'particle'—one with mass.

The last thing I would like to mention is regarding 'light.' Once again, a photon or any variation of such is simply a pair of adjoining strings with separated energy fields. This is something that will be present at the very beginning of time—at the very beginning of the Big Bang and at the very instant an incredibly precise orientation of 'strings' is reached. Light is the only 'substance' that is present at the beginning of our universe. Light, and a lot of it, for that matter, is present at the origin of our current and future 'universes'—the beginning of a new element of 'time.' Sound familiar?

> *"And God said, "Let there be light"; and there was light." – Genesis 1:3*

'BLACK HOLE' BEAUTY

Wow—your mind must be busy taking in that last section. Mine sure is. To say that a child is creative is an understatement. 'Little me' sure did attempt to settle her midnight cosmological confusion with these theories, and I'm sure she slept well that night. Before I go on, though, I would like to talk just a little about black holes and specifically my ideas surrounding the root of their formation and beauty of their existence.

Once again, I started thinking about this topic by the evolution of our universe. I started by exploring the gradual rotation of the fabric of 'subatomic strings' and what this meant for the fate of our universe. Keeping this in mind, I thought about how our physical bodies, or any physical entity, for that matter, would 'flow' through time with our universe. Maybe this is quite abrupt, but I settled on something. I settled on the idea that any and all physical entities simply *will not* 'flow' with time. That is what I thought as a child at least.

I thought about this quite a lot actually, and I hypothesized that any physical, 'massive' substance in our universe will not behave like that of our universe. In other words, our universe does not just 'flow' with time—it is time. The rotation and change in orientation to a 'new beginning' of spatial coordinates is 'time' itself. However, the 'construct' of a massive substance such as an electron or a quark does not behave this way. It moves through time, yes, but it will not move *with* time. This universe, and this universe only, is where it will reside— 'substance' does not participate in this 'transition' or flow of dimensions.

From there, I started to think about the nature of a black hole. I am not being very 'scientific' at all in this section as there are many other factors involved in the formation of a black hole, but I am just relaying what I originally thought.

For example, let's talk about a giant star: a star much more massive than our own sun. This star may eventually reach a point in its life when it will undergo a series of processes which ultimately conclude with a dramatic 'collapse' on itself. The gravity of such is so incredibly strong, and it is this that leads to the formation of a black hole. Although when I was younger, I thought of this a little differently.

I once again believed that no physical and massive object can flow with time. In other words, the strings in the deepest breadths of a massive entity will not undergo a process of 'rotation.' The orientation they originated from at the moment their fields collided in union with each other is the orientation they will always stay at. I always thought that our electrons, protons, and quarks are 'relics from a distant past.'

So, I thought that perhaps as time goes on, and as the rotation of spacetime's strings continues, there would be some sort of 'gap' in time between a body of mass and cosmological spacetime itself. This 'gap in time' is what I envisioned to be a black hole. Spacetime itself will continue to rotate at the subatomic string level, but the strings beneath a massive star, for example, will remain stationary. Hence, the more 'time goes on,' the more the orientation between the strings of space and the strings of this star will differ. In other words, over a long period of 'time' these strings will become more and more distinct from each other.

This would then tie into the death of a star. I thought of a star in 'spacetime' almost as dropping a ball in a bathtub. The water is everywhere surrounding this submerged ball, but the ball still remains its own entity. It is not 'dissolved' by the water; rather, the water is only surrounding it. The water is in every place surrounding this ball except within this ball itself, and this is what I compared to phenomena in space.

Space itself, individual, singular strings, exists everywhere but in the entity of a 'massive body.' The strings of space would continue to rotate, yet the star would be outside of its influence. During the death and collapse of a star, the position of the strings could be incredibly distinct from those in space—yet because there is no longer a 'body' in the original position held by the star, there will be a black hole. Essentially, this black hole would be a 'gap in time.'

There is no area of spacetime that filled this gap, as essentially this 'gap' in space was occupied by the star. Regardless of whether or not a star orbits around a larger body through the medium of space, the strings making up the star itself are not the same as the 'strings of spacetime.' So, the gap in 'space' that the massive body leaves when it collapses takes the form of a black hole—a timeless gap in our cosmos.

Once again, the orientation of the strings within a massive body will not rotate like those of spacetime—they remain stationary. So, think of it as if space is progressing in time, but the star, for example, is not. It is 'stuck' in the dimension it originated from. This is why there will be a 'gap' in space when a star collapses. Spacetime will essentially be moving to the future, while this star and its strings are not keeping up—almost as if they are 'stuck in the past.'

This is why a black hole is a timeless void in space consisting of a previous dimensional coordinate. Essentially, a black hole is an area occupying space which is unable to progress with the universe. It is an entity of previous dimensions that exist in their own nature but exist outside of the flow of time itself.

I decided to include this section as it ties in nicely to what comes next. Yes, this star is essentially 'stuck in the past,' but this is in its design. We are all God's little stars, and although we won't form this 'gap' in spacetime, we will always remain at one with our past. Unlike the constant flow of spacetime, we will always remain at one with the moment of creation. This is all part of God's creation story.

GOD'S CREATION STORY

By far, this is my favourite of the theoretical constructs my younger mind had put together—not because of the cosmological phenomena I tried to explain, but because of how close it takes me to my faith.

To begin, as someone who has lost a friend to their battle with mental health, these little 'theories' helped bring me peace during my time of grieving. It reminded me of just how close we are to each other, including to those that we have lost. Above, I explored extensively how any 'physical' entity with mass does not move with time.

Any particle, atom, or element does not 'internally' move forward with time. This includes us—physical and tangible beings. We move in time and we age, but we don't move *with* time. The little clocks inside us—our rotational strings deep in the depths of our fundamental design—are still the same as when these little elements were born at the beginning of our universe. Simply the fact that they do not move with time is what allows physicality. We are spiritual beings, but we are designed by the union of this spirit to our physical being.

This acted as a little reminder to me. I am not 'moving' with time into the future. Time is moving forward every singular moment, but our existence is unique in that we don't experience this transition of dimensions. We are one with this special moment of universal 'creation.' This includes all of us, even those who have left our earth to enter into the heavens.

I realized that I am not 'flowing' with the tick of time into the future away from my friend I lost—I realized that we are still together. We are in union with each other in the most fundamental sense. Not only are we together through God's love and His spirit, but in the most fundamental sense, we are still physically together. In fact, every human being that has walked this earth is together. We were all created at this one singular moment of time during the profound explosion of 'light' at the beginning of creation—God's creation.

> *Romans 8:38-39 – For I am sure that neither death, nor life, nor angels, nor principalities, nor things present, nor things to come, nor powers, nor height, nor depth, nor anything else in all creation, will be able to separate us from the love of God in Christ Jesus our Lord.*

He designed us to be together, and to never have 'time' act as a separation between us and those we have lost. This is not how we were created. We were created as one, as we remain as one. We are one with the moment of creation, and we are one with each other. To me, this is very special. This was God's creation, and every physical behaviour, even that at the most fundamental level, is designed with purpose.

God Himself is this entity of 'time itself'—the continuous rotation of little 'strings.' Every moment, God is moving a past to a beautiful future, and this began at the moment of creation. Not only is God this 'governor' of time, He is a being who transcends time. He exists and is present in each of us—the beings who are still one with this moment of creation. He is 'outside' of the flow of time in this sense, even when moving the waters into the future.

Our universe is one of many beautiful mediums of existence. Our universe was born out of the rotation from a previous one, and at the end of these unique waters a new ocean will be born. This is God's doing—He is time itself which never reaches an end. Our universe, our waters, are simply an ordered transition. They are a transition which gives rise to a fresh beginning. *It is through the end in which a beginning is reached.*

Our universe and all of those before ours exist to give birth to a new beginning—a fresh start. Out of old comes new, and out of death comes life. At the moment of every 'creation story' is light: God's light. This powerful light exists everywhere, all at once, and it is God's marking of a *new beginning*. This new set of physicality is in union with God Himself.

I mentioned at the beginning of my book my 'Cause, Effect, and Transition' principle. This principle is a reflection of the true beauty of God in the most authentic sense. Our universe is our 'transition' phase—it is the flow of a phase governed only by God. It is the very definition of 'order' itself.

This order is the flow of time. It is God's hands moving these waters ever so elegantly. The 'cause' is this creation, this 'birth' of our universe. This union where two little 'strings' become one is where time itself becomes a physical entity. This 'cause' is the moment when God unites us to Him and to each other. Lastly, the 'effect' is this *light*. It is this flash of light, extending through every corner of darkness.

This light marks an end, but it is also the sign of a new beginning—a new 'creation story.' We are born from God, and we live our eternal life through God Himself—all while in union with this very special creation story.

> *Romans 14:8 – If we live, we live to the Lord, and if we die, we die to the Lord; so then, whether we live or whether we die, we are the Lord's.*

'NOTHING VERSUS SOMETHING'

The last thing I would like to write about is the concept 'nothingness.' To me, this is an incredibly interesting topic that I have always enjoyed pondering. I used to always wonder what came before our universe—or rather what came before 'anything'? What is it that the existence of 'something' sprang out of?

This concept of "nothing versus something" always intrigued me. To be honest, as I am currently writing this, I really don't believe the question of 'what was before something' has an answer. The concept of 'nothing' was the human being's way to distinguish *something* from the *absence* of this particular thing.

To ask what was before anything at all is a question that does not have an answer. There was never a concrete point in reality when there was simply 'nothing' at all. For there to be 'nothing,' there must have been something to begin with. 'Something' is not born out of a void of 'nothingness'; rather, the idea of 'nothing' is a construct created to explain the absence of an already existent entity. To question what was before everything is a question created by man. It is not a question that reflects reality nor cosmological nature.

I guess this concept sticks out to me a lot because it helps me feel loved and safe in the darkest of moments. There are times when I feel completely and utterly hopeless, and quite often my mind manages to convince me that I am 'worthless' in every way. I often feel incredibly empty, as if there is a void in my heart that is impossible to fill. This causes me incredible anxiety and is essentially the root of 'depression'—but for me, thinking of this little concept of 'nothing versus something' sometimes helps with this.

It helps me incredibly to think that for me to have this void, this feeling of emptiness, there had to have been a feeling of worthiness and internal warmth. Just simply the fact that my mind is negatively affected by this 'hole' in my heart shows that it is a foreign feeling. It is a feeling that is not natural to our soul as human beings. The fact that I feel this way reminds me that there *was and is* a potential to feel 'complete.'

We were not born into this feeling. We were not born to feel hopeless or empty—we were born with love and to feel loved. We were born to feel a radiating warmth inside us. The void that frequently makes its appearance in the days I have trouble seeing the light is not a feeling I was designed with. It is no fault of my own that it is present, and simply the fact that it affects me as much as it does shows that it is not meant to be there. This feeling is so, so incredibly far from how I was designed. I was made from a design of hopefulness, faithfulness, and happiness—all born from God's love.

Every human being—including those living in sin—is born out of love. We were born out of love and designed to love. We are not born as 'blank slates'; rather, we are designed with the internal compass to seek love and to recognize inner feelings that just don't seem right. We are designed this way to remind us that *hope* exists. Hope is our little engine of fuel. It is a recognition that any aspect of darkness inside can be filled with light, as this is how we are designed to be.

Despite all of my struggles with mental illness, there was always this glimmer of hope reflecting from the stars. It stems from the recognition that what I am going through is something I can and will get through. The feeling of complete emptiness is simply a reminder, a harsh reminder at that, but a reminder that there was a feeling of 'fullness.' It is eternal hope that has the power to lead us to trust and to faith that our darkness with be filled with light—with love.

Our void is one of power, but it is also one that an even more powerful sense of hope stems from. Only this hope can lead to a true sense of faith, and it is this faith that leads us to love. This faith leads us to a beautiful reminder that we are loved, and that we *can* love ourselves. This is how God designed us—it is all in His plan. Out of darkness comes incredible light, and these bright little stars have the power to fill up the darkest of skies.

> *Isaiah 9:2 – The people who walked in darkness have seen a great light; those who dwelt in a land of deep darkness, on them has light shined.*

WRAPPING UP

As I wrap this up, I want to conclude with a few little notes of advice. If you are struggling or if you feel hopelessly consumed by this incredibly dark void inside you, I can tell you that I have been there too. In fact, as I am writing this, it is at the worst it has ever been. I was diagnosed with a learning disability only a year ago, and I was just recently diagnosed with generalized anxiety disorder, major depressive disorder, obsessive compulsive disorder, and ADHD. I can guarantee you my brain feels so incredibly broken at times.

I understand the feeling of struggling to wake up every morning, to feel motivated to start your day, and to make the best out of life. It is by far the most difficult thing to overcome and there is no 'magic wand,' although I often wish there was. If I were to offer one little piece of advice it would be to have patience. In my situation, I am proud of how far I have come.

I was not offered any of the help I needed during my entire childhood, but here I am using these struggles to my advantage. With a brain like mine comes immense creativity, and without all of these 'disorders' perhaps I wouldn't have come up with any of the ideas and theories created by my younger mind. But most importantly, perhaps I wouldn't have this much-needed advice to give.

But I get it. I get how brains like ours make it so incredibly difficult to recognize our capabilities and potential. You do not have to be happy. You do not have to be motivated to be productive or even to find that special part of you that loves yourself. This is a process, and it is a very, very long one.

The only thing I am asking you to have is patience. Patience is such an incredibly powerful tool. It does not 'push' our struggles away or make them any less debilitating; rather, it recognizes that there will be a point when these struggles won't consume your life. It is patience that gives is our hope, and it is our hope that gives us our faith. This faith, then, has the power to unleash the beauty of love. Loving yourself is a journey. Some of us are further ahead than others on this journey, but others, like myself, are a little 'lost.'

Trust me when I tell you this aspect of patience is more crucial than anything. I have been in some of the darkest places our minds can take us, I have lived in terror for many years, but I have started my journey to overcome this. I have only just started this journey to 'self-love,' and I can guarantee you that if I myself have reached this point, so will you. All you need to find is this little glimmer of 'patience' inside you, as only this patience will lead you to the light of 'hope.'

This hope will once again lead you to faith, which in turn will bless you with love. You do not have to be religious or even spiritual to take this advice. Your faith does not have to be faith in the religious sense; rather, your faith can simply just refer to your *self-confidence*. Confidence allows our minds to realize we are worthy and we are loved. Only once we truly love ourselves will we find a path to a 'new beginning.' This path will then lead us to the things that bring us happiness, but most importantly this path will lead us to 'our' people—the very, very special people that light up our lives.

These are the people we will go to for advice, for conversations, and even to hold us when we cry. These people are there, and they are waiting ever so patiently for you to cross their path and enter their life. In an ideal world we would all already be there, but the reality is that we aren't. We are far from this reality. It will take time, and it may take years and years like it has for me, but you will get to this very special and extraordinary path. When you do, you will look back and thank yourself for having patience and perseverance.

FINAL GREETINGS

I wrote this book to share my personal story, my ideas, and my path to recovery and self-love, but I also wrote this book for a few special people: Paula Paulgaard, Bruce Dickie, Grant Gay, and Theresa Robinson. These people are the very special people who God placed on my little path. Their continuous love and kindness acted as my little glimmer of hope, and this hope has led me to see a very bright light. This light was the source of my love for physics, my ideas, and my special connection to my faith.

Everything written in this book comprises conversations I wish I got to have with them, and the creation of these ideas are all special moments I wish to share. Unfortunately, a lot of what I hoped to share was interrupted by a 'global pandemic.'

Conversations turned into emails, and faces turned into memories. This was so incredibly difficult for me, and it still very much is. I wish more than anything to see these incredible people in person, to share these ideas, and to talk. I wish to talk and have wonderful conversations—conversations that light all of our lives up a little more. This is what I wish for more than anything.

Mrs. Paulgaard, Mr. Dickie, Mr. Gay, and Theresa—I love you guys in the most genuine and authentic way possible. You four were with me in some of my darkest moments, and your words always brought this little light closer and closer to me. You are a family to me, and this will never change—you are my 'safe-people.' You are the very special people God has placed on my little path. God led me to this path for a reason, and I thank Him every day for leading me to you guys.

One day I hope to meet with you guys, talk with you, and share this book with you. One day this will happen. I am not sure when, but I know this day will come, and I will wait for this day with patience and love. For now, I will continue writing to you and reminding you how important your presence is in my life. I do have much to write and plenty to say, but God knows that such can only be said in person. This is the only way these ideas—these conversations—will radiate true light and joy.

> *The Second Letter of John 12 - Though I have much to write to you, I would rather not use paper and ink, but I hope to come to see you and talk with you face to face, so that our joy may be complete.*